A Perspective on Single-Channel Frequency-Domain Speech Enhancement

Synthesis Lectures on Speech and Audio Processing

Editor
B.H. Juang, *Georgia Tech*

A Perspective on Single-Channel Frequency-Domain Speech Enhancement

Jacob Benesty and Yiteng Huang

ISBN: 978-3-031-01433-8   paperback
ISBN: 978-3-031-02561-7   ebook

DOI 10.1007/978-3-031-02561-7

A Publication in the Springer series
SYNTHESIS LECTURES ON SPEECH AND AUDIO PROCESSING
Lecture #8
Series Editor: B.H. Juang, *Georgia Tech*
Series ISSN
Synthesis Lectures on Speech and Audio Processing
Print 1932-121X   Electronic 1932-1678

A Perspective on Single-Channel Frequency-Domain Speech Enhancement

Jacob Benesty
INRS-EMT, University of Quebec

Yiteng Huang
WeVoice, Inc.

SYNTHESIS LECTURES ON SPEECH AND AUDIO PROCESSING #8

ABSTRACT

This book focuses on a class of single-channel noise reduction methods that are performed in the frequency domain via the short-time Fourier transform (STFT). The simplicity and relative effectiveness of this class of approaches make them the dominant choice in practical systems. Even though many popular algorithms have been proposed through more than four decades of continuous research, there are a number of critical areas where our understanding and capabilities still remain quite rudimentary, especially with respect to the relationship between noise reduction and speech distortion. All existing frequency-domain algorithms, no matter how they are developed, have one feature in common: the solution is eventually expressed as a gain function applied to the STFT of the noisy signal only in the current frame. As a result, the narrowband signal-to-noise ratio (SNR) cannot be improved, and any gains achieved in noise reduction on the fullband basis come with a price to pay, which is speech distortion. In this book, we present a new perspective on the problem by exploiting the difference between speech and typical noise in circularity and interframe self-correlation, which were ignored in the past. By gathering the STFT of the microphone signal of the current frame, its complex conjugate, and the STFTs in the previous frames, we construct several new, multiple-observation signal models similar to a microphone array system: there are multiple noisy speech observations, and their speech components are correlated but not completely coherent while their noise components are presumably uncorrelated. Therefore, the multichannel Wiener filter and the minimum variance distortionless response (MVDR) filter that were usually associated with microphone arrays will be developed for single-channel noise reduction in this book. This might instigate a paradigm shift geared toward speech distortionless noise reduction techniques.

KEYWORDS

single-channel noise reduction, speech enhancement, frequency domain, linear and widely linear models, interframe correlation, Wiener filter, maximum signal-to-noise ratio (SNR) filter, minimum variance distortionless response (MVDR) filter, tradeoff filter, linearly constrained minimum variance (LCMV) filter

Contents

CHAPTER 1

Introduction

Every speech communication and processing system suffers from the ubiquitous presence of additive noise. Today's widespread cellular phones and hands-free handsets are more likely to be used in acoustically adverse environments where background noise from different origins is loud and where the microphone may not be in close proximity to the speech source. The noise degrades the perceptual quality of speech and will impair the speech intelligibility when the signal-to-noise ratio (SNR) comes down to a certain level. Noise reduction intends to suppress such additive noise for purposes of speech enhancement. Operating on the noisy speech captured by a single microphone, noise reduction algorithms generally can enhance only the perceptual quality of speech when presented directly to a human listener with normal hearing but may improve both speech quality and intelligibility when the enhanced speech goes through a voice communication channel before being played out (30) and/or for the hearing impaired (12). So single-channel noise reduction (SCNR) has a large variety of applications including mobile phones, hearing aids, voice over Internet protocol (VoIP), just to name a few. Even after more than four decades of continuous research, SCNR remains a hot topic in speech and acoustic signal processing.

1.1 A BRIEF HISTORY OF SINGLE-CHANNEL NOISE REDUCTION (SCNR)

The first SCNR system was developed over 45 years ago by Schroeder (41), (42). The principle of Schroeder's system is the nowadays well-known spectral magnitude subtraction method. This work, however, has not received much public attention, probably because it is a purely analog implementation and, more importantly, it was never published in journals or conferences outside of the Bell System.

The interest in a digital form of the spectral subtraction technique was sparked by a 1974 paper by Weiss, Aschkenasy, and Parsons (47). A few years later, Boll, in his often-cited paper (11), reintroduced the spectral subtraction method yet for the first time in the framework of digital short-time Fourier analysis. These early algorithms were all based on an intuitive and simple idea: the clean speech spectrum can be restored by subtracting an estimate of the noise spectrum from the noisy speech spectrum, and the noise spectrum is estimated and updated during silent periods.

Though practically effective, the spectral magnitude subtraction approach is by no means optimal. It was thanks to the papers of (30) and (35) that the spectral subtraction technique began being examined in the framework of optimal estimation theory. This treatment initiated the development of many new noise reduction algorithms in the last three decades. They include the Wiener filter that intends to directly recover the *complex* (amplitude and phase) spectrum (i.e., the waveform in the time domain) of the clean speech (30), (35), and, in contrast, those in which only the spectral amplitude of the clean speech is estimated while its phase is copied from the phase of the noisy signal.

The spectral amplitude of a clean speech signal can be taken as the square root of a maximum likelihood (ML) estimate of its power spectrum. This leads to the spectral power subtraction method (10), (35), which is subtly different from the ML spectral amplitude estimator (35). In addition to the classical approach of ML estimation, the Bayesian decision rule was also found very useful. Ephraim and Malah introduced a celebrated minimum mean-square error (MMSE) estimator for spectral amplitude (MMSE-SA) in (18). This original idea was later enriched by the MMSE estimator for log spectral amplitude (MMSE-LSA) (19) and other generalized Bayesian estimators (31), (39), (49), which minimize the posterior expectation of various distance measures between the actual and estimated speech spectral amplitude. Maximum a posteriori (MAP) is another important Bayesian decision rule based on which Wolfe and Godsill developed a MAP spectral amplitude estimator (MAP-SA) (48).

In the aforementioned ML, MMSE, and MAP spectral amplitude estimators, it is commonly assumed that the short-time Fourier transforms (STFTs) of speech and noise are zero-mean, independent, complex Gaussian random processes. These assumptions are practically reasonable but may not be strictly true. Alternatively, a super-Gaussian model was suggested to be applied in combination with the MAP-SA approach in (33). More complicated statistical speech models (e.g., hidden Markov model) can also be used (20) but no close-form solution will be possibly deduced.

The development of SCNR algorithms is not necessarily restricted to the frequency domain. Actually it has been widely studied in the time domain and other transform domains too (see (8) and (9) for more complete discussions on those subjects). But the frequency-domain techniques are by far the most popular choice in practical systems for their simplicity and relative effectiveness.

1.2   RUDIMENTARY PROBLEM OF SCNR AND APPROACHES OF THIS BOOK

The existing frequency-domain SCNR algorithms differ in their optimization rules (ML, MMSE, or MAP), spectral distance measures (linear vs. log), and statistical models for speech (Gaussian, super-Gaussian, or HMM). But they all have one feature in common: the solution is eventually expressed as a (time-varying) gain function applied to the short-time Fourier transform (STFT) of the noisy signal in each frequency. Consequently, any gains in noise reduction by these algorithms always come at a price to pay, which is speech distortion.

This drawback of SCNR techniques is already well known. We believe that it is somehow due to a simplified formulation of the problem in which

1. the STFT coefficients of the speech and noise signals are all treated as second-order circular complex random variables, and

2. it has been implicitly assumed that the STFT of the current frame is uncorrelated with that in the neighboring frames even for the same frequency.

The first simplification corresponds to the misunderstanding that rotating the complex STFT coefficients of the microphone signal in the Argand plane or further exploring their complex conjugates will not offer any more helpful information for noise reduction. This unfortunately leads to only a partial utilization of the second order statistics of the signals. The second simplification is apparently not accurate for speech enhancement since speech is a highly self-correlated signal. Therefore, when we estimate the STFT of the clean speech in the current frame for one particular frequency bin, if we can take into account not only the STFT of the microphone signal in the same frame for the same frequency, but also its conjugate as well as the STFT coefficients in the neighboring frames, then we should be able to develop more sophisticated algorithms with hopefully better noise reduction results.

By gathering the STFT of the microphone signal of the current frame, its complex conjugate, and the STFTs in the previous frames, we can construct several new, multiple-observation signal models similar to a microphone array system: there are multiple noisy speech observations; their speech components are correlated but not completely coherent while their noise components are presumably uncorrelated. As a result, the multichannel Wiener filter and the minimum variance distortionless response (MVDR) filter that were usually associated with microphone arrays will be developed for SCNR in this book. This might instigate a paradigm shift geared toward speech distortionless noise reduction techniques.

1.3   ORGANIZATION OF THE BOOK

This book consists of 9 chapters including this introduction. Chapter 2 formulates the problem of SCNR and presents a general framework for studying the frequency-domain approaches. Chapter 3 describes some useful performance measures for SCNR without prescribing any specific forms for the noise reduction filter. The performance measures include input and output SNRs, gain in SNR, speech and noise reduction factors, and speech distortion index on both the narrowband and the fullband basis. In addition, the mean-square error (MSE) criterion is discussed and its relationships to the performance measures are explained. In Chapter 4, four linear and widely linear models for SCNR are introduced and discussed. Model 1 is the conventional model while the other three models are multiple-observation models. Model 2 exploits the noncircular characteristics of speech and includes the complex conjugates of the signal STFT coefficients. Model 3 considers the interframe correlation of speech and incorporates the STFT coefficients from both the current and the previous frames. Model 4 is the combination of Model 2 and Model 3, and so it is a more comprehensive and, meanwhile, a more complicated signal model. In Chapters 5 through 8, we develop and analyze the optimal noise reduction filters based on the four models, each of which is covered by a separate chapter. The discussed algorithms include the maximum SNR filter, the Wiener filter, the MVDR filter, the trade-off filter, and even the LCMV (linearly constrained minimum variance) filter if applicable. In Chapter 9, we evaluate the performance of the developed noise reduction filters with simulations.

CHAPTER 2

Problem Formulation

In this chapter, we formulate the problem of the additive noise picked up by a single microphone along with the desired signal (speech) in the time and frequency domains. We also propose a general framework for studying the frequency-domain speech enhancement problem.

2.1 SIGNAL MODEL

The noise reduction or speech enhancement problem considered in this study is one of recovering the desired signal (or clean speech) $x(t)$, t being the time index, of zero mean from the noisy observation (microphone signal) (4), (32), (44)

$$y(t) = x(t) + v(t), \tag{2.1}$$

where $v(t)$ is the unwanted additive noise, which is assumed to be a zero-mean random process white or colored but uncorrelated with $x(t)$. All signals are considered to be real and broadband.

Using the short-time Fourier transform (STFT), (2.1) can be rewritten in the frequency domain as

$$Y(k, m) = X(k, m) + V(k, m), \tag{2.2}$$

where $Y(k, m)$, $X(k, m)$, and $V(k, m)$ are the STFTs of $y(t)$, $x(t)$, and $v(t)$, respectively, at frequency-bin $k \in \{0, 1, \ldots, K - 1\}$ and time-frame m. Since $x(t)$ and $v(t)$ are uncorrelated by assumption, the variance of $Y(k, m)$ is

$$\begin{aligned}
\phi_Y(k, m) &= E\left[|Y(k, m)|^2\right] \\
&= \phi_X(k, m) + \phi_V(k, m),
\end{aligned} \tag{2.3}$$

where $E[\cdot]$ denotes mathematical expectation, and

$$\phi_X(k, m) = E\left[|X(k, m)|^2\right] \tag{2.4}$$

$$\phi_V(k, m) = E\left[|V(k, m)|^2\right] \tag{2.5}$$

are the variances of $X(k, m)$ and $V(k, m)$, respectively.

We define the pseudo-variance of $Y(k, m)$ as (23), (34), (36)

$$\begin{aligned}
\psi_Y(k, m) &= E\left[Y^2(k, m)\right] \\
&= \psi_X(k, m) + \psi_V(k, m),
\end{aligned} \tag{2.6}$$

where

$$\psi_X(k, m) = E\left[X^2(k, m)\right] \tag{2.7}$$

$$\psi_V(k, m) = E\left[V^2(k, m)\right] \tag{2.8}$$

are the pseudo-variances of $X(k, m)$ and $V(k, m)$, respectively. The pseudo-variance of a nonstationary signal (such as speech) in the frequency domain may not be equal to zero. Therefore, in order to have a complete picture of all second-order statistics in a noise reduction algorithm, the pseudo-variance needs to be taken into account (6), (7).

If we consider L consecutive time-frames of the observation signal, we can write (2.2) in a vector form as

$$\mathbf{y}(k, m) = \mathbf{x}(k, m) + \mathbf{v}(k, m), \tag{2.9}$$

where

$$\mathbf{y}(k, m) = \left[\begin{array}{cccc} Y(k, m) & Y(k, m-1) & \cdots & Y(k, m-L+1) \end{array}\right]^T$$

is a vector of length L, superscript ^{T} denotes transpose of a vector or a matrix, and $\mathbf{x}(k, m)$ and $\mathbf{v}(k, m)$ are defined in a similar way to $\mathbf{y}(k, m)$. The covariance matrix of $\mathbf{y}(k, m)$ is then

$$\Phi_{\mathbf{y}}(k, m) = E\left[\mathbf{y}(k, m)\mathbf{y}^H(k, m)\right]$$
$$= \Phi_{\mathbf{x}}(k, m) + \Phi_{\mathbf{v}}(k, m), \tag{2.10}$$

where the superscript ^{H} denotes transpose-conjugation, and

$$\Phi_{\mathbf{x}}(k, m) = E\left[\mathbf{x}(k, m)\mathbf{x}^H(k, m)\right] \tag{2.11}$$

$$\Phi_{\mathbf{v}}(k, m) = E\left[\mathbf{v}(k, m)\mathbf{v}^H(k, m)\right] \tag{2.12}$$

are the covariance matrices of $\mathbf{x}(k, m)$ and $\mathbf{v}(k, m)$, respectively.

The pseudo-covariance matrix of $\mathbf{y}(k, m)$ is defined as (23), (34), (36)

$$\Psi_{\mathbf{y}}(k, m) = E\left[\mathbf{y}(k, m)\mathbf{y}^T(k, m)\right]$$
$$= \Psi_{\mathbf{x}}(k, m) + \Psi_{\mathbf{v}}(k, m), \tag{2.13}$$

where

$$\Psi_{\mathbf{x}}(k, m) = E\left[\mathbf{x}(k, m)\mathbf{x}^T(k, m)\right] \tag{2.14}$$

$$\Psi_{\mathbf{v}}(k, m) = E\left[\mathbf{v}(k, m)\mathbf{v}^T(k, m)\right] \tag{2.15}$$

are the pseudo-covariance matrices of $\mathbf{x}(k, m)$ and $\mathbf{v}(k, m)$, respectively. For nonstationary signals, these pseudo-covariance matrices may be different from zero, so they should be exploited in the derivation of noise reduction filters.

2.2   GENERAL FRAMEWORK FOR SPEECH ENHANCEMENT

The objective of noise reduction/speech enhancement in the frequency domain via the STFT is to find a "good" estimate[1] of the desired signal $X(k, m)$ given $\mathbf{y}(k, m)$ and $\mathbf{y}^*(k, m)$, where the superscript $*$ denotes complex conjugation, with an appropriate function $f(\cdot)$, i.e.,

$$\widehat{X}(k, m) = f\left[\mathbf{y}(k, m), \mathbf{y}^*(k, m)\right] = f(\cdot). \tag{2.16}$$

In order to be able to define consistent performance measures for any function $f(\cdot)$, we need to decompose this function into two orthogonal components; one component that is completely coherent with the desired signal $X(k, m)$ and will, therefore, correspond to a linear function of the desired signal, and the other component that is incoherent with the desired signal and will, therefore, correspond to the residual interference-plus-noise. As a result, we can express (2.16) as

$$\begin{aligned}\widehat{X}(k, m) &= \rho_1^*(k, m)X(k, m) + U(k, m) \\ &= X_{\mathrm{ld}}(k, m) + U(k, m),\end{aligned} \tag{2.17}$$

where

$$X_{\mathrm{ld}}(k, m) = \rho_1^*(k, m)X(k, m) \tag{2.18}$$

is the linear version of the desired signal,

$$\rho_1(k, m) = \frac{E\left[X(k, m)\widehat{X}^*(k, m)\right]}{E\left[|X(k, m)|^2\right]} \tag{2.19}$$

is the normalized [with respect to $X(k, m)$] cross-correlation coefficient between $X(k, m)$ and $\widehat{X}(k, m)$,

$$U(k, m) = \widehat{X}(k, m) - \rho_1^*(k, m)X(k, m) \tag{2.20}$$

is the residual interference-plus-noise, and

$$E\left[X(k, m)U^*(k, m)\right] = 0. \tag{2.21}$$

Since the two components on the right-hand side of (2.17) are incoherent, the variance of $\widehat{X}(k, m)$ is

$$\begin{aligned}\phi_{\widehat{X}}(k, m) &= \phi_{X_{\mathrm{ld}}}(k, m) + \phi_U(k, m) \\ &= |\rho_1(k, m)|^2 \phi_X(k, m) + \phi_U(k, m),\end{aligned} \tag{2.22}$$

where

$$\begin{aligned}\phi_{X_{\mathrm{ld}}}(k, m) &= |\rho_1(k, m)|^2 \phi_X(k, m) \tag{2.23} \\ \phi_U(k, m) &= E\left[|U(k, m)|^2\right] \tag{2.24}\end{aligned}$$

---

[1]By "good" estimate, we mean that the additive noise is significantly reduced while the desired signal is lowly distorted or, ideally, not distorted.

are the variances of $X_{ld}(k, m)$ and $U(k, m)$, respectively. Clearly, the function $f(\cdot)$ must be derived in such a way that

$$|\rho_1(k, m)|^2 \rightarrow 1,$$
$$\phi_U(k, m) \rightarrow 0.$$

The term $|\rho_1(k, m)|^2$ contributes to the distortion of the speech signal [the closer is $|\rho_1(k, m)|^2$ to 1, the less the desired signal is distorted], while $\phi_U(k, m)$, which is the variance of the residual interference-plus-noise, plays a major role in the improvement in signal-to-noise ratio (SNR).

Continuing with this general concept, it is clear that the signal $U(k, m)$ contains a component that is completely coherent with the noise $V(k, m)$. Hence, we can also decompose $U(k, m)$ as follows:

$$\begin{aligned} U(k, m) &= \rho_2^*(k, m)V(k, m) + W(k, m) \\ &= V_{ln}(k, m) + W(k, m), \end{aligned} \tag{2.25}$$

where

$$\rho_2(k, m) = \frac{E\left[V(k, m)U^*(k, m)\right]}{E\left[|V(k, m)|^2\right]} \tag{2.26}$$

is the normalized [with respect to $V(k, m)$] cross-correlation coefficient between $V(k, m)$ and $U(k, m)$,

$$V_{ln}(k, m) = \rho_2^*(k, m)V(k, m) \tag{2.27}$$

is the linear component of the residual noise, $W(k, m)$ is the residual interference-plus-noise incoherent with $X(k, m)$ and $V(k, m)$, and

$$E\left[V(k, m)W^*(k, m)\right] = 0. \tag{2.28}$$

Substituting (2.25) into (2.17), the estimate of $X(k, m)$ becomes

$$\begin{aligned} \widehat{X}(k, m) &= \rho_1^*(k, m)X(k, m) + \rho_2^*(k, m)V(k, m) + W(k, m) \\ &= X_{ld}(k, m) + V_{ln}(k, m) + W(k, m). \end{aligned} \tag{2.29}$$

We observe that the estimate of the desired signal is the sum of three terms that are mutually incoherent. Therefore, the variance of $\widehat{X}(k, m)$ is

$$\begin{aligned} \phi_{\widehat{X}}(k, m) &= \phi_{X_{ld}}(k, m) + \phi_{V_{ln}}(k, m) + \phi_W(k, m) \\ &= \phi_{X_{ld}}(k, m) + \phi_U(k, m), \end{aligned} \tag{2.30}$$

where

$$\phi_{V_{ln}}(k, m) = |\rho_2(k, m)|^2 \phi_V(k, m), \tag{2.31}$$
$$\phi_W(k, m) = E\left[|W(k, m)|^2\right], \tag{2.32}$$
$$\phi_U(k, m) = \phi_{V_{ln}}(k, m) + \phi_W(k, m). \tag{2.33}$$

Here again, the function $f(\cdot)$ must be derived, if it is possible, in such a way that

$$
\begin{aligned}
|\rho_1(k,m)|^2 &\rightarrow 1, \\
|\rho_2(k,m)|^2 &\rightarrow 0, \\
\phi_W(k,m) &\rightarrow 0.
\end{aligned}
$$

C H A P T E R   3

Performance Measures

In this chapter, we are going to derive the most useful performance measures for speech enhancement in the general framework where any function $f(\cdot)$ can be used. We can divide these performance measures into two categories. The first category evaluates the noise reduction performance while the second one evaluates speech distortion. We also discuss the very convenient mean-square error (MSE) criterion and show how it is related to the performance measures.

3.1   NOISE REDUCTION

The signal-to-noise ratio (SNR) is the most important performance measure in the problem of speech enhancement. We have the input SNR (before processing) and the output SNR (after processing).

We define the narrowband and fullband input SNRs at time-frame m as (4)

$$\text{iSNR}(k, m) \quad = \quad \frac{\phi_X(k, m)}{\phi_V(k, m)}, \; k = 0, 1, \ldots, K - 1, \tag{3.1}$$

$$\text{iSNR}(m) \quad = \quad \frac{\sum_{k=0}^{K-1} \phi_X(k, m)}{\sum_{k=0}^{K-1} \phi_V(k, m)}. \tag{3.2}$$

It is easy to show that (4)

$$\text{iSNR}(m) \leq \sum_{k=0}^{K-1} \text{iSNR}(k, m). \tag{3.3}$$

To quantify the level of the interference-plus-noise remaining after the noise reduction processing via the function $f(\cdot)$, we define the narrowband output SNR as the ratio of the variance of the linear version of the desired signal over the variance of the residual interference-plus-noise[1], i.e.,

$$\text{oSNR}(k, m) \quad = \quad \frac{\phi_{X_{\text{ld}}}(k, m)}{\phi_{V_{\text{ln}}}(k, m) + \phi_W(k, m)}$$

$$= \quad \frac{|\rho_1(k, m)|^2 \, \phi_X(k, m)}{|\rho_2(k, m)|^2 \, \phi_V(k, m) + \phi_W(k, m)}, \; k = 0, 1, \ldots, K - 1. \tag{3.4}$$

---

[1]In this work, we consider the interference as part of the noise in the definitions of the performance measures.

We deduce that the narrowband gain in SNR is

$$
\begin{aligned}
\mathcal{A}(k, m) &= \frac{\text{oSNR}(k, m)}{\text{iSNR}(k, m)} \\
&= \frac{|\rho_1(k, m)|^2}{|\rho_2(k, m)|^2 + \text{RINR}(k, m)}, \quad k = 0, 1, \ldots, K - 1,
\end{aligned}
\tag{3.5}
$$

where

$$
\text{RINR}(k, m) = \frac{\phi_W(k, m)}{\phi_V(k, m)}, \quad k = 0, 1, \ldots, K - 1
\tag{3.6}
$$

is the residual-interference-plus-noise-to-noise ratio (RINR). We observe from (3.5) that $\mathcal{A}(k, m) \geq 1$ if and only if

$$
|\rho_1(k, m)|^2 \geq |\rho_2(k, m)|^2 + \text{RINR}(k, m).
\tag{3.7}
$$

We define the fullband output SNR at time-frame m as

$$
\text{oSNR}(m) = \frac{\sum_{k=0}^{K-1} |\rho_1(k, m)|^2 \phi_X(k, m)}{\sum_{k=0}^{K-1} \left[|\rho_2(k, m)|^2 \phi_V(k, m) + \phi_W(k, m) \right]}
\tag{3.8}
$$

and it can be verified that (4)

$$
\text{oSNR}(m) \leq \sum_{k=0}^{K-1} \text{oSNR}(k, m).
\tag{3.9}
$$

The previous inequality shows that the fullband output SNR is always upper bounded no matter the choice of $f(\cdot)$. The fullband gain in SNR at time-frame m is simply defined as

$$
\mathcal{A}(m) = \frac{\text{oSNR}(m)}{\text{iSNR}(m)}.
\tag{3.10}
$$

The noise reduction factor (3), (14) quantifies the amount of noise whose is rejected by the function $f(\cdot)$. This quantity is defined as the ratio of the variance of the noise at the microphone over the variance of the residual interference-plus-noise. The narrowband and fullband noise reduction factors are then

$$
\begin{aligned}
\xi_{\text{nr}}(k, m) &= \frac{\phi_V(k, m)}{\phi_{V_{\text{In}}}(k, m) + \phi_W(k, m)} \\
&= \frac{1}{|\rho_2(k, m)|^2 + \text{RINR}(k, m)} \\
&= \frac{\mathcal{A}(k, m)}{|\rho_1(k, m)|^2}, \quad k = 0, 1, \ldots, K - 1,
\end{aligned}
\tag{3.11}
$$

$$
\xi_{\text{nr}}(m) = \frac{\sum_{k=0}^{K-1} \phi_V(k, m)}{\sum_{k=0}^{K-1} \left[|\rho_2(k, m)|^2 \phi_V(k, m) + \phi_W(k, m) \right]},
\tag{3.12}
$$

and we always have

$$\xi_{nr}(m) \leq \sum_{k=0}^{K-1} \xi_{nr}(k, m). \tag{3.13}$$

The noise reduction factors are expected to be lower bounded by 1 for appropriate choices of $f(\cdot)$. So the more the noise is reduced, the higher are the values of the noise reduction factors.

3.2 SPEECH DISTORTION

In practice, the function $f(\cdot)$ may distort the desired signal. In order to evaluate the level of this distortion, we define the speech reduction factor (4) as the variance of the desired signal over the variance of the linear version of the desired signal. Therefore, the narrowband and fullband speech reduction factors are defined as

$$\begin{aligned} \xi_{sr}(k, m) &= \frac{\phi_X(k, m)}{\phi_{X_{ld}}(k, m)} \\ &= \frac{1}{|\rho_1(k, m)|^2}, \quad k = 0, 1, \ldots, K - 1, \end{aligned} \tag{3.14}$$

$$\xi_{sr}(m) = \frac{\sum_{k=0}^{K-1} \phi_X(k, m)}{\sum_{k=0}^{K-1} |\rho_1(k, m)|^2 \phi_X(k, m)}, \tag{3.15}$$

and we always have

$$\xi_{sr}(m) \leq \sum_{k=0}^{K-1} \xi_{sr}(k, m). \tag{3.16}$$

An important observation is that the choice of a function $f(\cdot)$ that does not distort the desired signal requires the constraint

$$|\rho_1(k, m)|^2 = 1, \quad \forall k, m. \tag{3.17}$$

Thus, the speech reduction factor is equal to 1 if there is no distortion and is expected to be greater than 1 when distortion occurs.

By making the appropriate substitutions, one can derive the relationships:

$$\begin{aligned} \mathcal{A}(k, m) &= \frac{\xi_{nr}(k, m)}{\xi_{sr}(k, m)}, \quad k = 0, 1, \ldots, K - 1, \end{aligned} \tag{3.18}$$

$$\mathcal{A}(m) = \frac{\xi_{nr}(m)}{\xi_{sr}(m)}. \tag{3.19}$$

When no distortion occurs, the gain in SNR coincides with the noise reduction factor.

Another useful performance measure is the speech distortion index (3), (14) defined as

$$
\begin{aligned}
\upsilon_{sd}(k, m) &= \frac{E\left\{|X_{ld}(k, m) - X(k, m)|^2\right\}}{\phi_X(k, m)} \\
&= \left|\rho_1^*(k, m) - 1\right|^2, \quad k = 0, 1, \ldots, K - 1
\end{aligned}
\tag{3.20}
$$

in the narrowband case and as

$$
\upsilon_{sd}(m) = \frac{\sum_{k=0}^{K-1} E\left\{|X_{ld}(k, m) - X(k, m)|^2\right\}}{\sum_{k=0}^{K-1} \phi_X(k, m)}
\tag{3.21}
$$

in the fullband case. The speech distortion index is always greater than or equal to 0 and should be upper bounded by 1 for appropriate choices of $f(\cdot)$; so the higher is its value, the more the desired signal is distorted.

3.3   MEAN-SQUARE ERROR (MSE) CRITERION

We define the error signal between the estimated and desired signals as

$$
\begin{aligned}
\mathcal{E}(k, m) &= \widehat{X}(k, m) - X(k, m) \\
&= X_{ld}(k, m) + V_{ln}(k, m) + W(k, m) - X(k, m),
\end{aligned}
\tag{3.22}
$$

which can be written as the sum of two incoherent error signals:

$$
\mathcal{E}(k, m) = \mathcal{E}_d(k, m) + \mathcal{E}_r(k, m),
\tag{3.23}
$$

where

$$
\begin{aligned}
\mathcal{E}_d(k, m) &= X_{ld}(k, m) - X(k, m) \\
&= \left[\rho_1^*(k, m) - 1\right] X(k, m)
\end{aligned}
\tag{3.24}
$$

is the signal distortion due to the function $f(\cdot)$ and

$$
\begin{aligned}
\mathcal{E}_r(k, m) &= V_{ln}(k, m) + W(k, m) \\
&= \rho_2^*(k, m) V(k, m) + W(k, m) \\
&= U(k, m)
\end{aligned}
\tag{3.25}
$$

represents the residual interference-plus-noise.

The narrowband mean-square error (MSE) criterion is then

$$
\begin{aligned}
J(k, m) &= E\left[|\mathcal{E}(k, m)|^2\right] \\
&= J_d(k, m) + J_r(k, m),
\end{aligned}
\tag{3.26}
$$

where

$$
\begin{aligned}
J_{\mathrm{d}}(k,m) &= E\left[|\mathcal{E}_{\mathrm{d}}(k,m)|^2\right] \\
&= E\left[|X_{\mathrm{ld}}(k,m) - X(k,m)|^2\right] \\
&= \left|\rho_1^*(k,m) - 1\right|^2 \phi_X(k,m)
\end{aligned} \tag{3.27}
$$

and

$$
\begin{aligned}
J_{\mathrm{r}}(k,m) &= E\left[|\mathcal{E}_{\mathrm{r}}(k,m)|^2\right] \\
&= |\rho_2(k,m)|^2 \phi_V(k,m) + \phi_W(k,m) \\
&= \phi_U(k,m).
\end{aligned} \tag{3.28}
$$

We define the first narrowband normalized MSE (NMSE) as

$$
\begin{aligned}
\widetilde{J}(k,m) &= \frac{J(k,m)}{\phi_V(k,m)} \\
&= \mathrm{iSNR}(k,m) \cdot \upsilon_{\mathrm{sd}}(k,m) + \frac{1}{\xi_{\mathrm{nr}}(k,m)} \\
&= \mathrm{iSNR}(k,m)\left[\upsilon_{\mathrm{sd}}(k,m) + \frac{1}{\mathrm{oSNR}(k,m) \cdot \xi_{\mathrm{sr}}(k,m)}\right],
\end{aligned} \tag{3.29}
$$

where we have used the following relationships

$$
\upsilon_{\mathrm{sd}}(k,m) = \frac{J_{\mathrm{d}}(k,m)}{\phi_X(k,m)}, \tag{3.30}
$$

$$
\xi_{\mathrm{nr}}(k,m) = \frac{\phi_V(k,m)}{J_{\mathrm{r}}(k,m)}. \tag{3.31}
$$

We define the second narrowband NMSE as

$$
\begin{aligned}
\overline{J}(k,m) &= \frac{J(k,m)}{\phi_X(k,m)} \\
&= \upsilon_{\mathrm{sd}}(k,m) + \frac{1}{\mathrm{oSNR}(k,m) \cdot \xi_{\mathrm{sr}}(k,m)}
\end{aligned} \tag{3.32}
$$

and, obviously,

$$
\widetilde{J}(k,m) = \mathrm{iSNR}(k,m) \cdot \overline{J}(k,m). \tag{3.33}
$$

We have shown how the NMSEs are related to the performance measures. In the same way, we define the fullband MSE at time-frame m as

$$
\begin{aligned}
J(m) &= \frac{1}{K}\sum_{k=0}^{K-1} J(k,m) \\
&= \frac{1}{K}\sum_{k=0}^{K-1} J_{\mathrm{d}}(k,m) + \frac{1}{K}\sum_{k=0}^{K-1} J_{\mathrm{r}}(k,m) \\
&= J_{\mathrm{d}}(m) + J_{\mathrm{r}}(m).
\end{aligned} \tag{3.34}
$$

We then deduce the fullband NMSEs at time-frame m:

$$\widetilde{J}(m) = K\frac{J(m)}{\sum_{k=0}^{K-1}\phi_V(k,m)}$$

$$= \text{iSNR}(m) \cdot \upsilon_{\text{sd}}(m) + \frac{1}{\xi_{\text{nr}}(m)}, \tag{3.35}$$

$$\overline{J}(m) = K\frac{J(m)}{\sum_{k=0}^{K-1}\phi_X(k,m)}$$

$$= \upsilon_{\text{sd}}(m) + \frac{1}{\text{oSNR}(m) \cdot \xi_{\text{sr}}(m)}. \tag{3.36}$$

It is straightforward to see that minimizing the narrowband MSE at each frequency-bin k is equivalent to minimizing the fullband MSE.

C H A P T E R 4

Linear and Widely Linear Models

From the general framework described in Chapter 2, the function $f(\cdot)$ can be nonlinear if we want to. But in this work, we limit our investigation to linear and widely linear models since they are the easiest ones to study and, arguably, the most practical ones to implement. In this chapter, we discuss four important models.

4.1 MODEL 1: CONVENTIONAL

In the simple conventional model, only a gain $H(k,m)$ is used to estimate the desired signal, $X(k,m)$, from the observation $Y(k,m)$, i.e.,

$$
\begin{aligned}
\widehat{X}(k,m) &= H(k,m)Y(k,m)\\
&= H(k,m)X(k,m) + H(k,m)V(k,m)\\
&= X_{\mathrm{ld}}(k,m) + V_{\mathrm{ln}}(k,m),\ k=0,1,\ldots,K-1,
\end{aligned}
\tag{4.1}
$$

where $H(k,m)$ is a complex gain factor whose module should be smaller than 1,

$$
X_{\mathrm{ld}}(k,m) = H(k,m)X(k,m)
\tag{4.2}
$$

is the linear version of the desired signal, and

$$
V_{\mathrm{ln}}(k,m) = H(k,m)V(k,m)
\tag{4.3}
$$

is the residual noise.

Identifying (4.1) with (2.29), it is obvious that for Model 1, we have

$$
\begin{aligned}
\rho_1^*(k,m) &= \rho_2^*(k,m) = H(k,m),\ \forall k,m,\\
W(k,m) &= 0,\ \forall k,m.
\end{aligned}
\tag{4.4}
\tag{4.5}
$$

The variance of $\widehat{X}(k,m)$ is

$$
\begin{aligned}
\phi_{\widehat{X}}(k,m) &= |H(k,m)|^2\phi_Y(k,m)\\
&= |H(k,m)|^2\phi_X(k,m) + |H(k,m)|^2\phi_V(k,m),\ k=0,1,\ldots,K-1.
\end{aligned}
\tag{4.6}
$$

Intuitively, we see from (4.6) that for the frequencies dominated by noise, the corresponding gains should be close to 0, while for the frequencies dominated by speech, the corresponding gains should be close to 1.

4.2   MODEL 2: WIDELY LINEAR

In the widely linear filtering approach (34), (38), two complex gain factors are applied to $Y(k, m)$ and $Y^*(k, m)$. This makes sense, since as long as $Y(k, m)$ and $Y^*(k, m)$ are not completely coherent [i.e., noncircular (37)], $Y^*(k, m)$ can be considered as another observation. Therefore, with Model 2, the estimate of $X(k, m)$ is

$$
\begin{aligned}
\widehat{X}(k, m) &= \widetilde{H}(k, m)Y(k, m) + \widetilde{H}'(k, m)Y^*(k, m) \\
&= \widetilde{\mathbf{h}}^{H}(k, m)\widetilde{\mathbf{y}}(k, m) \\
&= \widetilde{\mathbf{h}}^{H}(k, m)\widetilde{\mathbf{x}}(k, m) + \widetilde{\mathbf{h}}^{H}(k, m)\widetilde{\mathbf{v}}(k, m), \ k = 0, 1, \ldots, K - 1, \quad (4.7)
\end{aligned}
$$

where

$$
\begin{aligned}
\widetilde{\mathbf{h}}(k, m) &= \left[\begin{array}{c} \widetilde{H}^*(k, m) \\ \widetilde{H}'^*(k, m) \end{array} \right], \\
\widetilde{\mathbf{y}}(k, m) &= \left[\begin{array}{c} Y(k, m) \\ Y^*(k, m) \end{array} \right],
\end{aligned}
$$

and $\widetilde{\mathbf{x}}(k, m)$ and $\widetilde{\mathbf{v}}(k, m)$ are defined in a similar way to $\widetilde{\mathbf{y}}(k, m)$. If $\widetilde{H}'(k, m) = 0$ for any k and m, (4.7) degenerates to Model 1. This, however, will not happen in general for noncircular complex random variables (1), (2). We can rewrite (4.7) as

$$
\widehat{X}(k, m) = X_{\mathrm{f}}(k, m) + V_{\mathrm{rn}}(k, m), \ k = 0, 1, \ldots, K - 1, \quad (4.8)
$$

where

$$
\begin{aligned}
X_{\mathrm{f}}(k, m) &= \widetilde{\mathbf{h}}^{H}(k, m)\widetilde{\mathbf{x}}(k, m) \\
V_{\mathrm{rn}}(k, m) &= \widetilde{\mathbf{h}}^{H}(k, m)\widetilde{\mathbf{v}}(k, m)
\end{aligned}
$$

are, respectively, the filtered version of the desired signal and its complex conjugate, and the residual noise. From (4.8), we deduce the variance of $\widehat{X}(k, m)$:

$$
\phi_{\widehat{X}}(k, m) = \phi_{X_{\mathrm{f}}}(k, m) + \phi_{V_{\mathrm{rn}}}(k, m), \quad (4.9)
$$

where

$$
\begin{aligned}
\phi_{X_{\mathrm{f}}}(k, m) &= E\left[|X_{\mathrm{f}}(k, m)|^2 \right] \quad (4.10) \\
&= \widetilde{\mathbf{h}}^{H}(k, m)\Phi_{\widetilde{\mathbf{x}}}(k, m)\widetilde{\mathbf{h}}(k, m), \\
\phi_{V_{\mathrm{rn}}}(k, m) &= E\left[|V_{\mathrm{rn}}(k, m)|^2 \right] \quad (4.11) \\
&= \widetilde{\mathbf{h}}^{H}(k, m)\Phi_{\widetilde{\mathbf{v}}}(k, m)\widetilde{\mathbf{h}}(k, m),
\end{aligned}
$$

and

$$
\begin{aligned}
\Phi_{\widetilde{\mathbf{a}}}(k, m) &= E\left[\widetilde{\mathbf{a}}(k, m)\widetilde{\mathbf{a}}^H(k, m)\right] \\
&= \phi_A(k, m)\begin{bmatrix} 1 & \gamma_A(k, m) \\ \gamma_A^*(k, m) & 1 \end{bmatrix} \\
&= \phi_A(k, m)\Gamma_{\widetilde{\mathbf{a}}}(k, m)
\end{aligned}
\tag{4.12}
$$

is the covariance matrix of $\widetilde{\mathbf{a}}(k, m) = \begin{bmatrix} A(k, m) & A^*(k, m) \end{bmatrix}^T \in \{\widetilde{\mathbf{x}}(k, m), \widetilde{\mathbf{v}}(k, m)\}$ with

$$
\begin{aligned}
\gamma_A(k, m) &= \frac{\psi_A(k, m)}{\phi_A(k, m)} \\
&= \frac{E\left[A^2(k, m)\right]}{E\left[|A(k, m)|^2\right]}
\end{aligned}
\tag{4.13}
$$

being the (second-order) circularity quotient[1] (36) and

$$
\Gamma_{\widetilde{\mathbf{a}}}(k, m) = \begin{bmatrix} 1 & \gamma_A(k, m) \\ \gamma_A^*(k, m) & 1 \end{bmatrix}
\tag{4.14}
$$

being the circularity matrix. It can easily be shown that (36)

$$
0 \leq |\gamma_A(k, m)| \leq 1.
\tag{4.15}
$$

The circularity coefficient, $|\gamma_A(k, m)|$, conveys information about the degree of circularity of the signal $A(k, m)$. In particular, if $A(k, m)$ is a (second-order) circular complex random variable, then $\gamma_A(k, m) = 0$ and $\Gamma_{\widetilde{\mathbf{a}}}(k, m) = \mathbf{I}_2$, where

$$
\mathbf{I}_2 = \begin{bmatrix} 1 & 0 \\ 0 & 1 \end{bmatrix} = \begin{bmatrix} \mathbf{i}_1 & \mathbf{i}_2 \end{bmatrix}
\tag{4.16}
$$

is the 2×2 identity matrix.

The signal $X_f(k, m)$ consists of components from both the desired signal $X(k, m)$ and its conjugate. But not all these components are what we want. It is, therefore, necessary and important to distinguish between the filtered desired signal and the residual interference that both may exist in $X_f(k, m)$ at the same time. Specifically, $\widetilde{H}(k, m)X(k, m)$ is part of the overall filtered desired signal, but $\widetilde{H}'(k, m)X^*(k, m)$ is not. If $\gamma_X(k, m) = 0$ for any k and m, $X(k, m)$ and $X^*(k, m)$ are incoherent, and the overall filtered desired signal is indeed $\widetilde{H}(k, m)X(k, m)$. But for $\gamma_X(k, m) \neq 0$, $X^*(k, m)$ is coherent with $X(k, m)$ and contains both the desired signal and an interference component. Following the idea proposed in (16), we can decompose $X^*(k, m)$ into two orthogonal components:

$$
X^*(k, m) = \gamma_X^*(k, m)X(k, m) + X'(k, m),
\tag{4.17}
$$

---

[1]The circularity quotient of A is simply the ratio of the pseudo-variance of A over the variance of A or the coherence function between A and A^*.

where

$$X'(k, m) = X^*(k, m) - \gamma_X^*(k, m) X(k, m), \tag{4.18}$$

$$E\left[X(k, m) X'^*(k, m)\right] = 0, \tag{4.19}$$

and

$$\begin{aligned}
\phi_{X'}(k, m) &= E\left[\left|X'(k, m)\right|^2\right] \\
&= \phi_X(k, m)\left[1 - |\gamma_X(k, m)|^2\right].
\end{aligned} \tag{4.20}$$

We can then express (4.8) as

$$\begin{aligned}
\widehat{X}(k, m) &= X_{\mathrm{ld}}(k, m) + \widetilde{\mathbf{h}}^H(k, m)\mathbf{i}_2 X'(k, m) + V_{\mathrm{rn}}(k, m) \\
&= X_{\mathrm{ld}}(k, m) + U(k, m),
\end{aligned} \tag{4.21}$$

where

$$\begin{aligned}
X_{\mathrm{ld}}(k, m) &= \widetilde{\mathbf{h}}^H(k, m)\Gamma_{\widetilde{\mathbf{x}}}(k, m)\mathbf{i}_1 X(k, m) \\
&= \left[\widetilde{H}(k, m) + \gamma_X^*(k, m)\widetilde{H}'(k, m)\right] X(k, m)
\end{aligned} \tag{4.22}$$

$$\begin{aligned}
U(k, m) &= \widetilde{\mathbf{h}}^H(k, m)\mathbf{i}_2 X'(k, m) + V_{\mathrm{rn}}(k, m) \\
&= \widetilde{H}'(k, m) X'(k, m) + V_{\mathrm{rn}}(k, m)
\end{aligned} \tag{4.23}$$

are, respectively, the linear version of the desired signal and the residual interference-plus-noise.

We can also decompose $V^*(k, m)$ into two orthogonal components the same way we did it for $X^*(k, m)$. We get

$$V^*(k, m) = \gamma_V^*(k, m) V(k, m) + V'(k, m), \tag{4.24}$$

where $\gamma_V(k, m)$ and $V'(k, m)$ are defined in a similar way to $\gamma_X(k, m)$ and $X'(k, m)$. Now, (4.21) can be rewritten as

$$\widehat{X}(k, m) = X_{\mathrm{ld}}(k, m) + V_{\mathrm{ln}}(k, m) + \widetilde{H}'(k, m)\left[X'(k, m) + V'(k, m)\right], \tag{4.25}$$

where

$$\begin{aligned}
V_{\mathrm{ln}}(k, m) &= \widetilde{\mathbf{h}}^H(k, m)\Gamma_{\widetilde{\mathbf{v}}}(k, m)\mathbf{i}_1 V(k, m) \\
&= \left[\widetilde{H}(k, m) + \gamma_V^*(k, m)\widetilde{H}'(k, m)\right] V(k, m)
\end{aligned} \tag{4.26}$$

is the linear component of the residual noise. Comparing (4.25) with (2.29), we see that for Model 2, we have

$$\begin{aligned}
\rho_1^*(k, m) &= \widetilde{H}(k, m) + \gamma_X^*(k, m)\widetilde{H}'(k, m), \ \forall k, m, \tag{4.27} \\
\rho_2^*(k, m) &= \widetilde{H}(k, m) + \gamma_V^*(k, m)\widetilde{H}'(k, m), \ \forall k, m, \tag{4.28} \\
W(k, m) &= \widetilde{H}'(k, m)\left[X'(k, m) + V'(k, m)\right], \ \forall k, m. \tag{4.29}
\end{aligned}$$

The three terms on the right-hand side of (4.21) are mutually incoherent. Therefore, we have

$$\phi_{\widehat{X}}(k,m) = \phi_{X_{\mathrm{ld}}}(k,m) + \phi_U(k,m), \tag{4.30}$$

where

$$
\begin{aligned}
\phi_{X_{\mathrm{ld}}}(k,m) &= E\left[|X_{\mathrm{ld}}(k,m)|^2\right] \\
&= \phi_X(k,m)\left|\widetilde{\mathbf{h}}^H(k,m)\Gamma_{\widetilde{\mathbf{x}}}(k,m)\mathbf{i}_1\right|^2,
\end{aligned} \tag{4.31}
$$

$$
\begin{aligned}
\phi_U(k,m) &= E\left[|U(k,m)|^2\right] \\
&= \phi_{X'}(k,m)\left|\widetilde{\mathbf{h}}^H(k,m)\mathbf{i}_2\right|^2 + \phi_{V_{\mathrm{rn}}}(k,m).
\end{aligned} \tag{4.32}
$$

4.3   MODEL 3: INTERFRAME CORRELATION

In Model 3, the interframe correlation is taken into account. Hence, we estimate $X(k,m)$, $k = 0, 1, \ldots, K-1$, by passing $Y(k,m)$, $k = 0, 1, \ldots, K-1$, from consecutive time-frames through a finite-impulse-response-filter (FIR) of length L, i.e.,

$$
\begin{aligned}
\widehat{X}(k,m) &= \sum_{l=0}^{L-1} H_l(k,m)Y(k,m-l) \\
&= \mathbf{h}^H(k,m)\mathbf{y}(k,m), \quad k = 0, 1, \ldots, K-1,
\end{aligned} \tag{4.33}
$$

where L is the number of consecutive time-frames and

$$\mathbf{h}(k,m) = \begin{bmatrix} H_0^*(k,m) & H_1^*(k,m) & \cdots & H_{L-1}^*(k,m) \end{bmatrix}^T$$

is a vector of length L. The case $L = 1$ corresponds to the conventional frequency-domain approach (i.e., Model 1). Note that this concept, to take into account the interframe correlation in a speech enhancement algorithm, was introduced in (4), (5), (15) but in the Karhunen-Loève expansion (KLE) domain. In (17), the interframe correlation was also used to improve the a priori SNR estimator.

Let us now rewrite the signal $\widehat{X}(k,m)$ into the following form:

$$
\begin{aligned}
\widehat{X}(k,m) &= \mathbf{h}^H(k,m)\mathbf{x}(k,m) + \mathbf{h}^H(k,m)\mathbf{v}(k,m) \\
&= X_{\mathrm{f}}(k,m) + V_{\mathrm{rn}}(k,m), \quad k = 0, 1, \ldots, K-1,
\end{aligned} \tag{4.34}
$$

where

$$X_{\mathrm{f}}(k,m) = \mathbf{h}^H(k,m)\mathbf{x}(k,m) \tag{4.35}$$

is a filtered version of the desired signal at L consecutive time-frames and

$$V_{\mathrm{rn}}(k,m) = \mathbf{h}^H(k,m)\mathbf{v}(k,m) \tag{4.36}$$

is the residual noise which is uncorrelated with $X_f(k, m)$.

At time-frame m, our desired signal is $X(k, m)$ [and not the whole vector $\mathbf{x}(k, m)$]. However, the vector $\mathbf{x}(k, m)$ in $X_f(k, m)$ [eq. (4.34)] contains both the desired signal, $X(k, m)$, and the components $X(k, m - l)$, $l \neq 0$, which are not the desired signals at time-frame m but signals that are correlated with $X(k, m)$. Therefore, the elements $X(k, m - l)$, $l \neq 0$, contain both a part of the desired signal and a component that we consider as an interference. This suggests that we should decompose $X(k, m - l)$ into two orthogonal components, corresponding to the part of the desired signal and interference, i.e.,

$$X(k, m - l) = \rho_X^*(k, m, l)X(k, m) + X'(k, m - l), \tag{4.37}$$

where

$$X'(k, m - l) = X(k, m - l) - \rho_X^*(k, m, l)X(k, m), \tag{4.38}$$

$$E\left[X(k, m)X'^*(k, m - l)\right] = 0, \tag{4.39}$$

and

$$\rho_X(k, m, l) = \frac{E\left[X(k, m)X^*(k, m - l)\right]}{E\left[|X(k, m)|^2\right]} \tag{4.40}$$

is the interframe correlation coefficient of the signal $X(k, m)$. Hence, we can write the vector $\mathbf{x}(k, m)$ as

$$
\begin{aligned}
\mathbf{x}(k, m) &= X(k, m)\boldsymbol{\rho}_X^*(k, m) + \mathbf{x}'(k, m) \\
&= \mathbf{x}_d(k, m) + \mathbf{x}'(k, m),
\end{aligned}
\tag{4.41}
$$

where

$$\mathbf{x}_d(k, m) = X(k, m)\boldsymbol{\rho}_X^*(k, m) \tag{4.42}$$

is the desired signal vector,

$$\mathbf{x}'(k, m) = \left[\ X'(k, m)\quad X'(k, m - 1)\quad \cdots\quad X'(k, m - L + 1)\ \right]^T$$

is the interference signal vector, and

$$
\begin{aligned}
\boldsymbol{\rho}_X(k, m) &= \left[\ \rho_X(k, m, 0)\quad \rho_X(k, m, 1)\quad \cdots\quad \rho_X(k, m, L - 1)\ \right]^T \\
&= \left[\ 1\quad \rho_X(k, m, 1)\quad \cdots\quad \rho_X(k, m, L - 1)\ \right]^T \\
&= \frac{E\left[X(k, m)\mathbf{x}^*(k, m)\right]}{E\left[|X(k, m)|^2\right]}
\end{aligned}
\tag{4.43}
$$

is the (normalized) interframe correlation vector.

Substituting (4.41) into (4.34), we get

$$
\begin{aligned}
\widehat{X}(k, m) &= \mathbf{h}^H(k, m) \left[X(k, m) \boldsymbol{\rho}_X^*(k, m) + \mathbf{x}'(k, m) + \mathbf{v}(k, m) \right] \\
&= X_{\text{ld}}(k, m) + X_{\text{ri}}'(k, m) + V_{\text{rn}}(k, m),
\end{aligned}
\tag{4.44}
$$

where

$$
X_{\text{ld}}(k, m) = X(k, m) \mathbf{h}^H(k, m) \boldsymbol{\rho}_X^*(k, m)
\tag{4.45}
$$

is the linear version of the desired signal and

$$
X_{\text{ri}}'(k, m) = \mathbf{h}^H(k, m) \mathbf{x}'(k, m)
\tag{4.46}
$$

is the residual interference. We observe that the estimate of the desired signal is the sum of three terms that are mutually incoherent. The first one is clearly the filtered desired signal while the two others are the filtered undesired signals (interference-plus-noise). Therefore, the variance of $\widehat{X}(k, m)$ is

$$
\phi_{\widehat{X}}(k, m) = \phi_{X_{\text{ld}}}(k, m) + \phi_{X_{\text{ri}}'}(k, m) + \phi_{V_{\text{rn}}}(k, m),
\tag{4.47}
$$

where

$$
\begin{aligned}
\phi_{X_{\text{ld}}}(k, m) &= \phi_X(k, m) \left| \mathbf{h}^H(k, m) \boldsymbol{\rho}_X^*(k, m) \right|^2 \\
&= \mathbf{h}^H(k, m) \boldsymbol{\Phi}_{\mathbf{x}_{\text{d}}}(k, m) \mathbf{h}(k, m),
\end{aligned}
\tag{4.48}
$$

$$
\begin{aligned}
\phi_{X_{\text{ri}}'}(k, m) &= \mathbf{h}^H(k, m) \boldsymbol{\Phi}_{\mathbf{x}'}(k, m) \mathbf{h}(k, m) \\
&= \mathbf{h}^H(k, m) \boldsymbol{\Phi}_{\mathbf{x}}(k, m) \mathbf{h}(k, m) - \phi_X(k, m) \left| \mathbf{h}^H(k, m) \boldsymbol{\rho}_X^*(k, m) \right|^2,
\end{aligned}
\tag{4.49}
$$

$$
\phi_{V_{\text{rn}}}(k, m) = \mathbf{h}^H(k, m) \boldsymbol{\Phi}_{\mathbf{v}}(k, m) \mathbf{h}(k, m),
\tag{4.50}
$$

$$
\boldsymbol{\Phi}_{\mathbf{x}_{\text{d}}}(k, m) = \phi_X(k, m) \boldsymbol{\rho}_X^*(k, m) \boldsymbol{\rho}_X^T(k, m)
\tag{4.51}
$$

is the correlation matrix (whose rank is equal to 1) of $\mathbf{x}_{\text{d}}(k, m)$, and

$$
\boldsymbol{\Phi}_{\mathbf{z}}(k, m) = E\left[\mathbf{z}(k, m) \mathbf{z}^H(k, m) \right]
\tag{4.52}
$$

is the correlation matrix of $\mathbf{z}(k, m) \in \{\mathbf{x}(k, m), \mathbf{x}'(k, m), \mathbf{v}(k, m)\}$.

Let us now decompose $\mathbf{v}(k, m)$ as follows:

$$
\mathbf{v}(k, m) = V(k, m) \boldsymbol{\rho}_V^*(k, m) + \mathbf{v}'(k, m),
\tag{4.53}
$$

where $\boldsymbol{\rho}_V(k, m)$ and $\mathbf{v}'(k, m)$ are defined in a similar way to $\boldsymbol{\rho}_X(k, m)$ and $\mathbf{x}'(k, m)$. Expression (4.44) becomes

$$
\begin{aligned}
\widehat{X}(k, m) &= X(k, m) \mathbf{h}^H(k, m) \boldsymbol{\rho}_X^*(k, m) + V(k, m) \mathbf{h}^H(k, m) \boldsymbol{\rho}_V^*(k, m) \\
&\quad + \mathbf{h}^H \left[\mathbf{x}'(k, m) + \mathbf{v}'(k, m) \right].
\end{aligned}
\tag{4.54}
$$

Identifying (4.54) with (2.29), we observe that for Model 3, we have

$$
\begin{aligned}
\rho_1^*(k, m) &= \mathbf{h}^H(k, m)\boldsymbol{\rho}_X^*(k, m), \; \forall k, m, & (4.55) \\
\rho_2^*(k, m) &= \mathbf{h}^H(k, m)\boldsymbol{\rho}_V^*(k, m), \; \forall k, m, & (4.56) \\
W(k, m) &= \mathbf{h}^H \left[\mathbf{x}'(k, m) + \mathbf{v}'(k, m) \right], \; \forall k, m. & (4.57)
\end{aligned}
$$

4.4   MODEL 4: WIDELY LINEAR AND INTERFRAME CORRELATION

In our fourth and last model, we use the widely linear filtering approach (34), (38), and take into account the interframe correlation. In this scenario, the estimate of the desired signal is

$$
\begin{aligned}
\widehat{X}(k, m) &= \widetilde{\mathbf{h}}^H(k, m)\mathbf{y}(k, m) + \widetilde{\mathbf{h}}'^H(k, m)\mathbf{y}^*(k, m) \\
&= \widetilde{\underline{\mathbf{h}}}^H(k, m)\widetilde{\underline{\mathbf{y}}}(k, m), \; k = 0, 1, \ldots, K - 1, \quad (4.58)
\end{aligned}
$$

where $\widetilde{\mathbf{h}}(k, m)$ and $\widetilde{\mathbf{h}}'(k, m)$ are two complex FIR filters of length L, and

$$
\begin{aligned}
\widetilde{\underline{\mathbf{h}}}(k, m) &= \left[\begin{array}{c} \widetilde{\mathbf{h}}(k, m) \\ \widetilde{\mathbf{h}}'(k, m) \end{array} \right], \\
\widetilde{\underline{\mathbf{y}}}(k, m) &= \left[\begin{array}{c} \mathbf{y}(k, m) \\ \mathbf{y}^*(k, m) \end{array} \right].
\end{aligned}
$$

Expression (4.58) can be rewritten as

$$
\begin{aligned}
\widehat{X}(k, m) &= \widetilde{\underline{\mathbf{h}}}^H(k, m)\widetilde{\underline{\mathbf{x}}}(k, m) + \widetilde{\underline{\mathbf{h}}}^H(k, m)\widetilde{\underline{\mathbf{v}}}(k, m) \\
&= X_{\mathrm{f}}(k, m) + V_{\mathrm{rn}}(k, m), \; k = 0, 1, \ldots, K - 1, \quad (4.59)
\end{aligned}
$$

where $\widetilde{\underline{\mathbf{x}}}(k, m)$ and $\widetilde{\underline{\mathbf{v}}}(k, m)$ are defined in a similar way to $\widetilde{\underline{\mathbf{y}}}(k, m)$, and

$$
\begin{aligned}
X_{\mathrm{f}}(k, m) &= \widetilde{\underline{\mathbf{h}}}^H(k, m)\widetilde{\underline{\mathbf{x}}}(k, m) \\
V_{\mathrm{rn}}(k, m) &= \widetilde{\underline{\mathbf{h}}}^H(k, m)\widetilde{\underline{\mathbf{v}}}(k, m)
\end{aligned}
$$

are, respectively, the filtered version of $\widetilde{\underline{\mathbf{x}}}(k, m)$ and the residual noise. From (4.59), we deduce the variance of $\widehat{X}(k, m)$:

$$
\phi_{\widehat{X}}(k, m) = \phi_{X_{\mathrm{f}}}(k, m) + \phi_{V_{\mathrm{rn}}}(k, m), \quad (4.60)
$$

where

$$
\begin{aligned}
\phi_{X_{\mathrm{f}}}(k, m) &= E\left[|X_{\mathrm{f}}(k, m)|^2 \right] & (4.61) \\
&= \widetilde{\underline{\mathbf{h}}}^H(k, m)\Phi_{\widetilde{\underline{\mathbf{x}}}}(k, m)\widetilde{\underline{\mathbf{h}}}(k, m), \\
\phi_{V_{\mathrm{rn}}}(k, m) &= E\left[|V_{\mathrm{rn}}(k, m)|^2 \right] & (4.62) \\
&= \widetilde{\underline{\mathbf{h}}}^H(k, m)\Phi_{\widetilde{\underline{\mathbf{v}}}}(k, m)\widetilde{\underline{\mathbf{h}}}(k, m),
\end{aligned}
$$

and

$$\Phi_{\widetilde{\underline{\mathbf{a}}}}(k, m) = E\left[\widetilde{\underline{\mathbf{a}}}(k, m)\widetilde{\underline{\mathbf{a}}}^H(k, m)\right]$$

$$= \begin{bmatrix} \Phi_{\mathbf{a}}(k, m) & \Psi_{\mathbf{a}}(k, m) \\ \Psi_{\mathbf{a}}^*(k, m) & \Phi_{\mathbf{a}}^*(k, m) \end{bmatrix} \qquad (4.63)$$

is the covariance matrix of $\widetilde{\underline{\mathbf{a}}}(k, m) = \begin{bmatrix} \mathbf{a}^T(k, m) & \mathbf{a}^H(k, m) \end{bmatrix}^T \in \left\{\widetilde{\underline{\mathbf{x}}}(k, m), \widetilde{\underline{\mathbf{v}}}(k, m)\right\}$.

In order to extract the desired signal at time-frame m, as we already did for Model 2 and Model 3, we need to decompose the vector $\widetilde{\underline{\mathbf{x}}}(k, m)$ into two orthogonal components:

$$\widetilde{\underline{\mathbf{x}}}(k, m) = X(k, m)\boldsymbol{\varrho}_X^*(k, m) + \widetilde{\underline{\mathbf{x}}}'(k, m)$$

$$= \mathbf{x}_{\mathrm{d}}(k, m) + \widetilde{\underline{\mathbf{x}}}'(k, m), \qquad (4.64)$$

where

$$\mathbf{x}_{\mathrm{d}}(k, m) = X(k, m)\boldsymbol{\varrho}_X^*(k, m) \qquad (4.65)$$

is the desired signal vector, $\widetilde{\underline{\mathbf{x}}}'(k, m)$ is the interference signal vector,

$$\boldsymbol{\varrho}_X(k, m) = \begin{bmatrix} \varrho_X(k, m, 0) & \varrho_X(k, m, 1) & \cdots & \varrho_X(k, m, 2L - 1) \end{bmatrix}^T$$

$$= \begin{bmatrix} 1 & \varrho_X(k, m, 1) & \cdots & \varrho_X(k, m, 2L - 1) \end{bmatrix}^T$$

$$= \frac{E\left[X(k, m)\widetilde{\underline{\mathbf{x}}}^*(k, m)\right]}{E\left[|X(k, m)|^2\right]} \qquad (4.66)$$

is the concatenation of the (normalized) interframe correlation and the (normalized) pseudo-correlation vectors, and

$$\varrho_X(k, m, l) = \begin{cases} \dfrac{E\left[X(k, m)X^*(k, m - l)\right]}{E\left[|X(k, m)|^2\right]}, & 0 \le l \le L - 1 \\ \dfrac{E\left[X(k, m)X(k, m - l)\right]}{E\left[|X(k, m)|^2\right]}, & L \le l \le 2L - 1 \end{cases} . \qquad (4.67)$$

Substituting (4.64) into (4.59), we obtain

$$\widehat{X}(k, m) = \widetilde{\underline{\mathbf{h}}}^H(k, m)\left[X(k, m)\boldsymbol{\varrho}_X^*(k, m) + \widetilde{\underline{\mathbf{x}}}'(k, m) + \widetilde{\underline{\mathbf{v}}}(k, m)\right]$$

$$= X_{\mathrm{ld}}(k, m) + X_{\mathrm{ri}}'(k, m) + V_{\mathrm{rn}}(k, m), \qquad (4.68)$$

where

$$X_{\mathrm{ld}}(k, m) = X(k, m)\widetilde{\underline{\mathbf{h}}}^H(k, m)\boldsymbol{\varrho}_X^*(k, m) \qquad (4.69)$$

is the linear version of the desired signal and

$$X_{\mathrm{ri}}'(k, m) = \widetilde{\underline{\mathbf{h}}}^H(k, m)\widetilde{\underline{\mathbf{x}}}'(k, m) \qquad (4.70)$$

is the residual interference. The variance of $\widehat{X}(k, m)$ is then

$$\phi_{\widehat{X}}(k, m) = \phi_{X_{\mathrm{ld}}}(k, m) + \phi_{X'_{\mathrm{ri}}}(k, m) + \phi_{V_{\mathrm{rn}}}(k, m), \tag{4.71}$$

where

$$
\begin{aligned}
\phi_{X_{\mathrm{ld}}}(k, m) &= \phi_X(k, m) \left| \widetilde{\underline{\mathbf{h}}}^H(k, m) \boldsymbol{\varrho}_X^*(k, m) \right|^2 \\
&= \widetilde{\underline{\mathbf{h}}}^H(k, m) \Phi_{\mathbf{x}_{\mathrm{d}}}(k, m) \widetilde{\underline{\mathbf{h}}}(k, m), \tag{4.72} \\
\phi_{X'_{\mathrm{ri}}}(k, m) &= \widetilde{\underline{\mathbf{h}}}^H(k, m) \Phi_{\widetilde{\underline{\mathbf{x}}}'}(k, m) \widetilde{\underline{\mathbf{h}}}(k, m) \\
&= \widetilde{\underline{\mathbf{h}}}^H(k, m) \Phi_{\widetilde{\underline{\mathbf{x}}}}(k, m) \widetilde{\underline{\mathbf{h}}}(k, m) - \phi_X(k, m) \left| \widetilde{\underline{\mathbf{h}}}^H(k, m) \boldsymbol{\varrho}_X^*(k, m) \right|^2, \tag{4.73} \\
\phi_{V_{\mathrm{rn}}}(k, m) &= \widetilde{\underline{\mathbf{h}}}^H(k, m) \Phi_{\widetilde{\underline{\mathbf{v}}}}(k, m) \widetilde{\underline{\mathbf{h}}}(k, m), \tag{4.74}
\end{aligned}
$$

$$\Phi_{\mathbf{x}_{\mathrm{d}}}(k, m) = \phi_X(k, m) \boldsymbol{\varrho}_X^*(k, m) \boldsymbol{\varrho}_X^T(k, m) \tag{4.75}$$

is the correlation matrix (whose rank is equal to 1) of $\mathbf{x}_{\mathrm{d}}(k, m)$, and

$$\Phi_{\widetilde{\underline{\mathbf{x}}}'}(k, m) = E\left[\widetilde{\underline{\mathbf{x}}}'(k, m) \widetilde{\underline{\mathbf{x}}}'^H(k, m) \right] \tag{4.76}$$

is the correlation matrix of $\widetilde{\underline{\mathbf{x}}}'(k, m)$.

By decomposing

$$\widetilde{\underline{\mathbf{v}}}(k, m) = V(k, m) \boldsymbol{\varrho}_V^*(k, m) + \widetilde{\underline{\mathbf{v}}}'(k, m), \tag{4.77}$$

where $\boldsymbol{\varrho}_V(k, m)$ and $\widetilde{\underline{\mathbf{v}}}'(k, m)$ are defined in a similar way to $\boldsymbol{\varrho}_X(k, m)$ and $\widetilde{\underline{\mathbf{x}}}'(k, m)$, (4.68) becomes

$$
\begin{aligned}
\widehat{X}(k, m) &= X(k, m) \widetilde{\underline{\mathbf{h}}}^H(k, m) \boldsymbol{\varrho}_X^*(k, m) + V(k, m) \widetilde{\underline{\mathbf{h}}}^H(k, m) \boldsymbol{\varrho}_V^*(k, m) \\
&\quad + \widetilde{\underline{\mathbf{h}}}^H \left[\widetilde{\underline{\mathbf{x}}}'(k, m) + \widetilde{\underline{\mathbf{v}}}'(k, m) \right]. \tag{4.78}
\end{aligned}
$$

Comparing (4.78) with (2.29), we observe that for Model 4, we have

$$
\begin{aligned}
\rho_1^*(k, m) &= \widetilde{\underline{\mathbf{h}}}^H(k, m) \boldsymbol{\varrho}_X^*(k, m), \ \forall k, m, \tag{4.79} \\
\rho_2^*(k, m) &= \widetilde{\underline{\mathbf{h}}}^H(k, m) \boldsymbol{\varrho}_V^*(k, m), \ \forall k, m, \tag{4.80} \\
W(k, m) &= \widetilde{\underline{\mathbf{h}}}^H \left[\widetilde{\underline{\mathbf{x}}}'(k, m) + \widetilde{\underline{\mathbf{v}}}'(k, m) \right], \ \forall k, m. \tag{4.81}
\end{aligned}
$$

CHAPTER 5

Optimal Filters with Model 1

In this chapter, we study noise reduction with Model 1. This is the simplest model and the most investigated one in the literature. Here, only an appropriate gain, $H(k, m)$, is used at each frequency-bin k in order to enhance the quality of the speech signal corrupted by an additive noise.

5.1 PERFORMANCE MEASURES

This section gives the performance measures with Model 1. They are easy to derive from Chapter 3 and Section 4.1, Chapter 4.

The narrowband and fullband input SNRs are already defined in (3.1) and (3.2). They are independent of the model we use.

We easily deduce the narrowband output SNR

$$
\begin{aligned}
\text{oSNR}[H(k, m)] &= \frac{|H(k, m)|^2 \phi_X(k, m)}{|H(k, m)|^2 \phi_V(k, m)} \\
&= \frac{\phi_X(k, m)}{\phi_V(k, m)}, \quad k = 0, 1, \dots, K - 1
\end{aligned}
\tag{5.1}
$$

and the fullband output SNR

$$
\text{oSNR}[H(:, m)] = \frac{\sum_{k=0}^{K-1} |H(k, m)|^2 \phi_X(k, m)}{\sum_{k=0}^{K-1} |H(k, m)|^2 \phi_V(k, m)}.
\tag{5.2}
$$

We notice that the narrowband output SNR, which is equal to the narrowband input SNR, cannot be improved with just a gain, but the fullband output SNR can.

The narrowband and fullband noise reduction factors are

$$
\begin{aligned}
\xi_{\text{nr}}[H(k, m)] &= \frac{\phi_V(k, m)}{|H(k, m)|^2 \phi_V(k, m)} \\
&= \frac{1}{|H(k, m)|^2}, \quad k = 0, 1, \dots, K - 1,
\end{aligned}
\tag{5.3}
$$

$$
\begin{aligned}
\xi_{\text{nr}}[H(:, m)] &= \frac{\sum_{k=0}^{K-1} \phi_V(k, m)}{\sum_{k=0}^{K-1} |H(k, m)|^2 \phi_V(k, m)} \\
&= \frac{\sum_{k=0}^{K-1} \phi_V(k, m)}{\sum_{k=0}^{K-1} \xi_{\text{nr}}^{-1}[H(k, m)] \phi_V(k, m)}.
\end{aligned}
\tag{5.4}
$$

The noise reduction factor is supposed to have a lower bound of 1 for optimal gains, and the larger its value, the more the noise is reduced.

To quantify the speech distortion, we give the narrowband speech distortion index

$$
\begin{aligned}
\upsilon_{sd}\left[H(k,m)\right] &= \frac{E\left\{\left|H(k,m)X(k,m) - X(k,m)\right|^2\right\}}{\phi_X(k,m)} \\
&= \left|H(k,m) - 1\right|^2, \quad k = 0, 1, \ldots, K-1
\end{aligned}
\tag{5.5}
$$

and the fullband speech distortion index

$$
\begin{aligned}
\upsilon_{sd}\left[H(:,m)\right] &= \frac{\sum_{k=0}^{K-1} E\left\{\left|H(k,m)X(k,m) - X(k,m)\right|^2\right\}}{\sum_{k=0}^{K-1} \phi_X(k,m)} \\
&= \frac{\sum_{k=0}^{K-1} \upsilon_{sd}\left[H(k,m)\right]\phi_X(k,m)}{\sum_{k=0}^{K-1} \phi_X(k,m)}.
\end{aligned}
\tag{5.6}
$$

The speech distortion index is usually upper bounded by 1 for optimal gains.

Another way to quantify signal distortion is via the speech reduction factor. The narrowband and fullband definitions are

$$
\begin{aligned}
\xi_{sr}\left[H(k,m)\right] &= \frac{\phi_X(k,m)}{\left|H(k,m)\right|^2 \phi_X(k,m)} \\
&= \frac{1}{\left|H(k,m)\right|^2}, \quad k = 0, 1, \ldots, K-1,
\end{aligned}
\tag{5.7}
$$

$$
\begin{aligned}
\xi_{sr}\left[H(:,m)\right] &= \frac{\sum_{k=0}^{K-1} \phi_X(k,m)}{\sum_{k=0}^{K-1} \left|H(k,m)\right|^2 \phi_X(k,m)} \\
&= \frac{\sum_{k=0}^{K-1} \phi_X(k,m)}{\sum_{k=0}^{K-1} \xi_{sr}^{-1}\left[H(k,m)\right]\phi_X(k,m)}.
\end{aligned}
\tag{5.8}
$$

The speech reduction factor is supposed to have a lower bound of 1 for optimal gains.

In the frequency domain and with Model 1, the error signal between the estimated and desired signals at the frequency-bin k is

$$
\begin{aligned}
\mathcal{E}(k,m) &= \widehat{X}(k,m) - X(k,m) \\
&= H(k,m)Y(k,m) - X(k,m),
\end{aligned}
\tag{5.9}
$$

which can also be written as the sum of two incoherent error signals:

$$
\mathcal{E}(k,m) = \mathcal{E}_d(k,m) + \mathcal{E}_r(k,m),
\tag{5.10}
$$

where

$$
\mathcal{E}_d(k,m) = H(k,m)X(k,m) - X(k,m)
\tag{5.11}
$$

is the speech distortion due to the gain and

$$\mathcal{E}_r(k, m) = H(k, m)V(k, m) \tag{5.12}$$

represents the residual noise.

From the error signal (5.9), we give the corresponding frequency-domain (or narrowband) MSE criterion:

$$
\begin{aligned}
J[H(k, m)] &= E\left[|\mathcal{E}(k, m)|^2\right] \\
&= \phi_X(k, m) + |H(k, m)|^2 \phi_Y(k, m) - 2\mathcal{R}[H(k, m)\phi_{YX}(k, m)] \\
&= \phi_X(k, m) + |H(k, m)|^2 \phi_Y(k, m) - 2\mathcal{R}[H(k, m)\phi_X(k, m)],
\end{aligned} \tag{5.13}
$$

where $\mathcal{R}(\cdot)$ is the real part of a complex number and

$$
\begin{aligned}
\phi_{YX}(k, m) &= E\left[Y(k, m)X^*(k, m)\right] \\
&= \phi_X(k, m)
\end{aligned}
$$

is the cross-correlation between the signals $Y(k, m)$ and $X(k, m)$. Expression (5.13) can be structured in a different way:

$$
\begin{aligned}
J[H(k, m)] &= E\left[|\mathcal{E}_d(k, m)|^2\right] + E\left[|\mathcal{E}_r(k, m)|^2\right] \\
&= J_d[H(k, m)] + J_r[H(k, m)].
\end{aligned} \tag{5.14}
$$

The most interesting optimal gains can be derived from the previous narrowband MSEs.

5.2   WIENER GAIN

By minimizing $J[H(k, m)]$ [eq. (5.13)] with respect to $H(k, m)$, we easily find the Wiener gain

$$
\begin{aligned}
H_W(k, m) &= \frac{E\left[|X(k, m)|^2\right]}{E\left[|Y(k, m)|^2\right]} \\
&= 1 - \frac{E\left[|V(k, m)|^2\right]}{E\left[|Y(k, m)|^2\right]} \\
&= \frac{\phi_X(k, m)}{\phi_X(k, m) + \phi_V(k, m)} \\
&= \frac{\text{iSNR}(k, m)}{1 + \text{iSNR}(k, m)}.
\end{aligned} \tag{5.15}
$$

We see that the noncausal Wiener gain is always real and positive. Furthermore, $0 \leq H_W(k, m) \leq 1$, $\forall k, m$.

We deduce the different narrowband performance measures:

$$\widetilde{J}\left[H_{\mathrm{W}}(k,m)\right] = \frac{\mathrm{iSNR}(k,m)}{1+\mathrm{iSNR}(k,m)} \leq 1, \tag{5.16}$$

$$\xi_{\mathrm{nr}}\left[H_{\mathrm{W}}(k,m)\right] = \left[1+\frac{1}{\mathrm{iSNR}(k,m)}\right]^2 \geq 1 \tag{5.17}$$

$$= \xi_{\mathrm{sr}}\left[H_{\mathrm{W}}(k,m)\right],$$

$$\upsilon_{\mathrm{sd}}\left[H_{\mathrm{W}}(k,m)\right] = \frac{1}{\left[1+\mathrm{iSNR}(k,m)\right]^2} \leq 1. \tag{5.18}$$

The fullband output SNR is

$$\mathrm{oSNR}\left[H_{\mathrm{W}}(:,m)\right] = \frac{\sum_{k=0}^{K-1}\phi_X(k,m)\left[\dfrac{\mathrm{iSNR}(k,m)}{1+\mathrm{iSNR}(k,m)}\right]^2}{\sum_{k=0}^{K-1}\phi_V(k,m)\left[\dfrac{\mathrm{iSNR}(k,m)}{1+\mathrm{iSNR}(k,m)}\right]^2}. \tag{5.19}$$

Property 5.1 With the optimal frequency-domain Wiener gain given in (5.15), the fullband output SNR is always greater than or equal to the input SNR, i.e., $\mathrm{oSNR}\left[H_{\mathrm{W}}(:,m)\right] \geq \mathrm{iSNR}(m)$.

Proof. We can use exactly the same techniques as the ones exposed in (4) to show this property. $\square$

Property 5.2 We have

$$\frac{\mathrm{iSNR}(m)}{1+\mathrm{oSNR}\left[H_{\mathrm{W}}(:,m)\right]} \leq \widetilde{J}\left[H_{\mathrm{W}}(:,m)\right] \leq \frac{\mathrm{iSNR}(m)}{1+\mathrm{iSNR}(m)}, \tag{5.20}$$

$$\frac{\left\{1+\mathrm{oSNR}\left[H_{\mathrm{W}}(:,m)\right]\right\}^2}{\mathrm{iSNR}(m)\cdot\mathrm{oSNR}\left[H_{\mathrm{W}}(:,m)\right]} \leq \xi_{\mathrm{nr}}\left[H_{\mathrm{W}}(:,m)\right] \leq \frac{\left[1+\mathrm{iSNR}(m)\right]\left\{1+\mathrm{oSNR}\left[H_{\mathrm{W}}(:,m)\right]\right\}}{\mathrm{iSNR}^2(m)}, \tag{5.21}$$

$$\frac{1}{\left\{1+\mathrm{oSNR}\left[H_{\mathrm{W}}(:,m)\right]\right\}^2} \leq \upsilon_{\mathrm{sd}}\left[H_{\mathrm{W}}(:,m)\right] \leq \frac{1+\mathrm{oSNR}\left[H_{\mathrm{W}}(:,m)\right]-\mathrm{iSNR}(m)}{\left[1+\mathrm{iSNR}(m)\right]\left\{1+\mathrm{oSNR}\left[H_{\mathrm{W}}(:,m)\right]\right\}}. \tag{5.22}$$

Proof. We can use exactly the same techniques as the ones exposed in (4) to show these different inequalities. $\square$

5.3   TRADEOFF GAIN

The tradeoff gain is obtained by minimizing the speech distortion with the constraint that the residual noise level is equal to a value smaller than the level of the original noise. This is equivalent to solving the problem

$$\min_{H(k,m)} J_d[H(k,m)] \quad \text{subject to} \quad J_r[H(k,m)] = \beta\phi_V(k,m), \tag{5.23}$$

where

$$\begin{aligned} J_d[H(k,m)] &= |1 - H(k,m)|^2 \phi_X(k,m), &(5.24)\\ J_r[H(k,m)] &= |H(k,m)|^2 \phi_V(k,m), &(5.25) \end{aligned}$$

and $0 < \beta < 1$ in order to have some noise reduction at the frequency-bin k. If we use a Lagrange multiplier, $\mu \geq 0$, to adjoin the constraint to the cost function, we get the tradeoff gain

$$\begin{aligned} H_{\mathrm{T},\mu}(k,m) &= \frac{\phi_X(k,m)}{\phi_X(k,m) + \mu\phi_V(k,m)} \\ &= \frac{\phi_Y(k,m) - \phi_V(k,m)}{\phi_Y(k,m) + (\mu-1)\phi_V(k,m)} \\ &= \frac{\mathrm{iSNR}(k,m)}{\mu + \mathrm{iSNR}(k,m)}. \end{aligned} \tag{5.26}$$

This gain can be seen as a frequency-domain Wiener gain with adjustable input noise level $\mu\phi_V(k,m)$. The particular cases of $\mu = 1$ and $\mu = 0$ correspond to the Wiener and distortionless gains, respectively.

The fullband output SNR is

$$\mathrm{oSNR}\left[H_{\mathrm{T},\mu}(:,m)\right] = \frac{\sum_{k=0}^{K-1} \phi_X(k,m)\left[\dfrac{\mathrm{iSNR}(k,m)}{\mu + \mathrm{iSNR}(k,m)}\right]^2}{\sum_{k=0}^{K-1} \phi_V(k,m)\left[\dfrac{\mathrm{iSNR}(k,m)}{\mu + \mathrm{iSNR}(k,m)}\right]^2}. \tag{5.27}$$

Property 5.3   With the tradeoff gain given in (5.26), the fullband output SNR is always greater than or equal to the input SNR, i.e., $\mathrm{oSNR}\left[H_{\mathrm{T},\mu}(:,m)\right] \geq \mathrm{iSNR}(m), \ \forall\mu \geq 0$.

Proof. We can use exactly the same techniques as the ones exposed in (4) to show this property. $\square$

From (5.27), we deduce that

$$\lim_{\mu\to\infty} \mathrm{oSNR}\left[H_{\mathrm{T},\mu}(:,m)\right] = \frac{\sum_{k=0}^{K-1} \phi_X(k,m)\mathrm{iSNR}^2(k,m)}{\sum_{k=0}^{K-1} \phi_V(k,m)\mathrm{iSNR}^2(k,m)} \leq \sum_{k=0}^{K-1} \mathrm{iSNR}(k,m). \tag{5.28}$$

This shows how the fullband output SNR of the tradeoff gain is upper bounded.

The fullband speech distortion index is

$$
\upsilon_{\mathrm{sd}}\left[H_{\mathrm{T},\mu}(:,m)\right] = \frac{\sum_{k=0}^{K-1} \dfrac{\mu^2 \phi_X(k,m)}{\left[\mu + \mathrm{iSNR}(k,m)\right]^2}}{\sum_{k=0}^{K-1} \phi_X(k,m)}.
\tag{5.29}
$$

Property 5.4   The fullband speech distortion index of the tradeoff gain is an increasing function of the parameter μ.

Proof.   It is straightforward to verify that

$$
\frac{d\upsilon_{\mathrm{sd}}\left[H_{\mathrm{T},\mu}(:,m)\right]}{d\mu} \geq 0,
\tag{5.30}
$$

which ends the proof.   □

It is clear that

$$
0 \leq \upsilon_{\mathrm{sd}}\left[H_{\mathrm{T},\mu}(:,m)\right] \leq 1, \ \forall\mu \geq 0.
\tag{5.31}
$$

Therefore, as μ increases, the fullband output SNR increases at the price of more distortion to the desired signal.

The tradeoff gain can be more general if we make the factor β dependent on the frequency, i.e., $\beta(k)$. By doing so, the control between noise reduction and speech distortion can be more effective since each frequency-bin k can be controlled independently of the others. With this consideration, we can easily see that the optimal gain derived from the criterion (5.23) is now

$$
H_{\mathrm{T},\mu}(k,m) = \frac{\mathrm{iSNR}(k,m)}{\mu(k) + \mathrm{iSNR}(k,m)},
\tag{5.32}
$$

where $\mu(k)$ is the frequency-dependent Lagrange multiplier. This approach can now provide some noise spectral shaping for masking by the speech signal (21), (25), (26), (28), (45), (46).

5.4   MAXIMUM SIGNAL-TO-NOISE RATIO (SNR) FILTER

Let us define the $K \times 1$ vector

$$
\mathbf{h}(m) = \left[\ H(0,m) \quad H(1,m) \quad \cdots \quad H(K-1,m)\ \right]^T,
\tag{5.33}
$$

which contains all the narrowband gains. The fullband output SNR can be rewritten as

$$
\begin{aligned}
\text{oSNR}\left[H(:,m)\right] &= \text{oSNR}\left[\mathbf{h}(m)\right] \\
&= \frac{\mathbf{h}^H(m)\mathbf{D}_{\phi_X}(m)\mathbf{h}(m)}{\mathbf{h}^H(m)\mathbf{D}_{\phi_V}(m)\mathbf{h}(m)},
\end{aligned} \tag{5.34}
$$

where

$$
\begin{aligned}
\mathbf{D}_{\phi_X}(m) &= \text{diag}\left[\phi_X(0,m),\phi_X(1,m),\ldots,\phi_X(K-1,m)\right] \tag{5.35} \\
\mathbf{D}_{\phi_V}(m) &= \text{diag}\left[\phi_V(0,m),\phi_V(1,m),\ldots,\phi_V(K-1,m)\right] \tag{5.36}
\end{aligned}
$$

are two diagonal matrices. We assume here that $\phi_V(k,m) \neq 0$, $\forall k, m$.

In the maximum SNR approach, we find the filter, $\mathbf{h}(m)$, that maximizes the fullband output SNR defined in (5.34). The solution to this problem that we denote by $\mathbf{h}_{\max}(m)$ is simply the eigenvector corresponding to the maximum eigenvalue of the matrix $\mathbf{D}_{\phi_V}^{-1}(m)\mathbf{D}_{\phi_X}(m)$. Since this matrix is diagonal, its maximum eigenvalue is its largest diagonal element, i.e.,

$$
\max_k \frac{\phi_X(k,m)}{\phi_V(k,m)} = \max_k \text{iSNR}(k,m). \tag{5.37}
$$

Assume that this maximum is the k_0th diagonal element of the matrix $\mathbf{D}_{\phi_V}^{-1}(m)\mathbf{D}_{\phi_X}(m)$. In this case, the k_0th component of $\mathbf{h}_{\max}(m)$ is 1, and all its other components are 0. As a result,

$$
\begin{aligned}
\text{oSNR}\left[\mathbf{h}_{\max}(m)\right] &= \max_k \text{iSNR}(k,m) \\
&= \text{iSNR}(k_0,m). \tag{5.38}
\end{aligned}
$$

We also deduce that

$$
\text{oSNR}\left[\mathbf{h}(m)\right] \leq \max_k \text{iSNR}(k,m), \ \forall \mathbf{h}(m). \tag{5.39}
$$

This means that with the Wiener, tradeoff, or any other gain, the fullband output SNR cannot exceed the maximum narrowband input SNR, which is a very interesting result on its own.

It is easy to derive the fullband speech distortion index:

$$
\upsilon_{\text{sd}}\left[\mathbf{h}_{\max}(m)\right] = 1 - \frac{\phi_X(k_0,m)}{\sum_{k=0}^{K-1}\phi_X(k,m)}, \tag{5.40}
$$

which can be very close to 1, implying very large distortions of the desired signal.

Needless to say that this maximum SNR filter is never used in practice since all narrowband signals but one are suppressed. But this filter is still interesting from a theoretical point of view.

CHAPTER 6

Optimal Filters with Model 2

This chapter is dedicated to the derivation and study of optimal filters with Model 2 where the conjugate of the noisy signal, i.e., $Y^*(k, m)$, is considered as another observation since, in general, it is not completely coherent with $Y(k, m)$. Therefore, we need to use the widely linear filtering technique (34), (38).

6.1 PERFORMANCE MEASURES

We should be able to derive the most important performance measures from Chapter 3 and Section 4.2, Chapter 4.

The narrowband output SNR is defined as

$$
\begin{aligned}
\mathrm{oSNR}\left[\widetilde{\mathbf{h}}(k, m)\right] &= \frac{\phi_{X_{\mathrm{ld}}}(k, m)}{\phi_U(k, m)} \\
&= \frac{\phi_X(k, m)\left|\widetilde{\mathbf{h}}^H(k, m)\Gamma_{\widetilde{\mathbf{x}}}(k, m)\mathbf{i}_1\right|^2}{\widetilde{\mathbf{h}}^H(k, m)\Phi_{\mathrm{in}}(k, m)\widetilde{\mathbf{h}}(k, m)}, \quad k = 0, 1, \ldots, K-1, \quad (6.1)
\end{aligned}
$$

where

$$
\Phi_{\mathrm{in}}(k, m) = \phi_{X'}(k, m)\mathbf{i}_2\mathbf{i}_2^T + \Phi_{\widetilde{\mathbf{v}}}(k, m) \tag{6.2}
$$

is the interference-plus-noise covariance matrix. For the particular filter $\widetilde{\mathbf{h}}(k, m) = \mathbf{i}_1 = \begin{bmatrix} 1 & 0 \end{bmatrix}^T$ (identity filter), we have

$$
\mathrm{oSNR}\left[\mathbf{i}_1(k, m)\right] = \mathrm{iSNR}(k, m). \tag{6.3}
$$

And for the filter $\widetilde{\mathbf{h}}(k, m) = \begin{bmatrix} H^*(k, m) & 0 \end{bmatrix}^T$ (equivalent to Model 1), we also have

$$
\mathrm{oSNR}\left[H^*(k, m)\right] = \mathrm{iSNR}(k, m). \tag{6.4}
$$

In the two previous scenarios, the narrowband SNR cannot be improved.

For any two vectors $\widetilde{\mathbf{h}}(k, m)$ and $\Gamma_{\widetilde{\mathbf{x}}}(k, m)\mathbf{i}_1$, and a positive definite matrix $\Phi_{\mathrm{in}}(k, m)$, we have

$$
\begin{aligned}
\left|\widetilde{\mathbf{h}}^H(k, m)\Gamma_{\widetilde{\mathbf{x}}}(k, m)\mathbf{i}_1\right|^2 &\leq \\
&\left[\widetilde{\mathbf{h}}^H(k, m)\Phi_{\mathrm{in}}(k, m)\widetilde{\mathbf{h}}(k, m)\right]\left[\mathbf{i}_1^T\Gamma_{\widetilde{\mathbf{x}}}(k, m)\Phi_{\mathrm{in}}^{-1}(k, m)\Gamma_{\widetilde{\mathbf{x}}}(k, m)\mathbf{i}_1\right].
\end{aligned} \tag{6.5}
$$

Using the previous inequality in (6.1), we deduce an upper bound for the narrowband output SNR:

$$\mathrm{oSNR}\left[\widetilde{\mathbf{h}}(k, m)\right] \leq \phi_X(k, m)\mathbf{i}_1^T \Gamma_{\widetilde{\mathbf{x}}}(k, m)\Phi_{\mathrm{in}}^{-1}(k, m)\Gamma_{\widetilde{\mathbf{x}}}(k, m)\mathbf{i}_1, \ k = 0, 1, \ldots, K-1. \quad (6.6)$$

We define the fullband output SNR at time-frame m as

$$\mathrm{oSNR}\left[\widetilde{\mathbf{h}}(:, m)\right] = \frac{\sum_{k=0}^{K-1} \phi_X(k, m) \left|\widetilde{\mathbf{h}}^H(k, m)\Gamma_{\widetilde{\mathbf{x}}}(k, m)\mathbf{i}_1\right|^2}{\sum_{k=0}^{K-1} \widetilde{\mathbf{h}}^H(k, m)\Phi_{\mathrm{in}}(k, m)\widetilde{\mathbf{h}}(k, m)} \quad (6.7)$$

and we always have

$$\mathrm{oSNR}\left[\widetilde{\mathbf{h}}(:, m)\right] \leq \sum_{k=0}^{K-1} \mathrm{oSNR}\left[\widetilde{\mathbf{h}}(k, m)\right] \leq$$

$$\sum_{k=0}^{K-1} \phi_X(k, m)\mathbf{i}_1^T \Gamma_{\widetilde{\mathbf{x}}}(k, m)\Phi_{\mathrm{in}}^{-1}(k, m)\Gamma_{\widetilde{\mathbf{x}}}(k, m)\mathbf{i}_1. \quad (6.8)$$

The previous inequality shows that the fullband output SNR is upper bounded no matter the $\widetilde{\mathbf{h}}(k, m)$, $k = 0, 1, \ldots, K-1$.

The narrowband and fullband noise reduction factors are

$$\xi_{\mathrm{nr}}\left[\widetilde{\mathbf{h}}(k, m)\right] = \frac{\phi_V(k, m)}{\phi_U(k, m)}$$

$$= \frac{\phi_V(k, m)}{\widetilde{\mathbf{h}}^H(k, m)\Phi_{\mathrm{in}}(k, m)\widetilde{\mathbf{h}}(k, m)}, \ k = 0, 1, \ldots, K-1, \quad (6.9)$$

$$\xi_{\mathrm{nr}}\left[\widetilde{\mathbf{h}}(:, m)\right] = \frac{\sum_{k=0}^{K-1} \phi_V(k, m)}{\sum_{k=0}^{K-1} \widetilde{\mathbf{h}}^H(k, m)\Phi_{\mathrm{in}}(k, m)\widetilde{\mathbf{h}}(k, m)}. \quad (6.10)$$

In the same manner, we define the narrowband and fullband speech reduction factors as

$$\xi_{\mathrm{sr}}\left[\widetilde{\mathbf{h}}(k, m)\right] = \frac{\phi_X(k, m)}{\phi_X(k, m)\left|\widetilde{\mathbf{h}}^H(k, m)\Gamma_{\widetilde{\mathbf{x}}}(k, m)\mathbf{i}_1\right|^2}$$

$$= \frac{1}{\left|\widetilde{\mathbf{h}}^H(k, m)\Gamma_{\widetilde{\mathbf{x}}}(k, m)\mathbf{i}_1\right|^2}, \ k = 0, 1, \ldots, K-1, \quad (6.11)$$

$$\xi_{\mathrm{sr}}\left[\widetilde{\mathbf{h}}(:, m)\right] = \frac{\sum_{k=0}^{K-1} \phi_X(k, m)}{\sum_{k=0}^{K-1} \phi_X(k, m)\left|\widetilde{\mathbf{h}}^H(k, m)\Gamma_{\widetilde{\mathbf{x}}}(k, m)\mathbf{i}_1\right|^2}. \quad (6.12)$$

Another way to quantify the speech distortion is via the narrowband speech distortion index

$$\upsilon_{\mathrm{sd}}\left[\widetilde{\mathbf{h}}(k, m)\right] = \frac{E\left\{|X_{\mathrm{ld}}(k, m) - X(k, m)|^2\right\}}{\phi_X(k, m)}$$

$$= \left|\widetilde{\mathbf{h}}^H(k, m)\Gamma_{\widetilde{\mathbf{x}}}(k, m)\mathbf{i}_1 - 1\right|^2, \ k = 0, 1, \ldots, K-1 \quad (6.13)$$

and the fullband speech distortion index

$$\upsilon_{\text{sd}}\left[\widetilde{\mathbf{h}}(:,m)\right] = \frac{\sum_{k=0}^{K-1} E\left\{|X_{\text{ld}}(k,m) - X(k,m)|^2\right\}}{\sum_{k=0}^{K-1} \phi_X(k,m)}. \tag{6.14}$$

A key observation from (6.11) or (6.13) is that the design of a noise reduction algorithm that does not distort the desired signal requires the constraint

$$\widetilde{\mathbf{h}}^H(k,m)\Gamma_{\widetilde{\mathbf{x}}}(k,m)\mathbf{i}_1 = 1. \tag{6.15}$$

Let us now give the error signal between the estimated and desired signals at the frequency-bin k for Model 2:

$$\begin{aligned}
\mathcal{E}(k,m) &= \widehat{X}(k,m) - X(k,m) \\
&= \widetilde{\mathbf{h}}^H(k,m)\widetilde{\mathbf{y}}(k,m) - X(k,m).
\end{aligned} \tag{6.16}$$

We can rewrite (6.16) as

$$\mathcal{E}(k,m) = \mathcal{E}_{\text{d}}(k,m) + \mathcal{E}_{\text{r}}(k,m), \tag{6.17}$$

where

$$\begin{aligned}
\mathcal{E}_{\text{d}}(k,m) &= X_{\text{ld}}(k,m) - X(k,m) \\
&= \left[\widetilde{\mathbf{h}}^H(k,m)\Gamma_{\widetilde{\mathbf{x}}}(k,m)\mathbf{i}_1 - 1\right] X(k,m)
\end{aligned} \tag{6.18}$$

is the speech distortion due to the complex filter, and

$$\begin{aligned}
\mathcal{E}_{\text{r}}(k,m) &= U(k,m) \\
&= \widetilde{\mathbf{h}}^H(k,m)\mathbf{i}_2 X'(k,m) + \widetilde{\mathbf{h}}^H(k,m)\widetilde{\mathbf{v}}(k,m)
\end{aligned} \tag{6.19}$$

represents the residual interference-plus-noise.

From the error signal (6.16), we can form the corresponding frequency-domain MSE criterion:

$$\begin{aligned}
J\left[\widetilde{\mathbf{h}}(k,m)\right] &= E\left[|\mathcal{E}(k,m)|^2\right] \\
&= E\left[|\mathcal{E}_{\text{d}}(k,m)|^2\right] + E\left[|\mathcal{E}_{\text{r}}(k,m)|^2\right] \\
&= J_{\text{d}}\left[\widetilde{\mathbf{h}}(k,m)\right] + J_{\text{r}}\left[\widetilde{\mathbf{h}}(k,m)\right].
\end{aligned} \tag{6.20}$$

Depending on how we optimize the different MSEs with Model 2, we obtain different optimal filters.

6.2   MAXIMUM SNR FILTER

The maximum SNR filter, $\widetilde{\mathbf{h}}_{\max}(k, m)$, is obtained by maximizing the narrowband output SNR as defined in (6.1). Therefore, $\widetilde{\mathbf{h}}_{\max}(k, m)$ is the eigenvector corresponding to the maximum eigenvalue of the matrix

$$\phi_X(k, m)\Phi_{\text{in}}^{-1}(k, m)\Gamma_{\widetilde{\mathbf{x}}}(k, m)\mathbf{i}_1\mathbf{i}_1^T\Gamma_{\widetilde{\mathbf{x}}}(k, m).$$

Let us denote this eigenvalue by $\widetilde{\lambda}_{\max}(k, m)$. Since the rank of the matrix $\Gamma_{\widetilde{\mathbf{x}}}(k, m)\mathbf{i}_1\mathbf{i}_1^T\Gamma_{\widetilde{\mathbf{x}}}(k, m)$ is equal to 1, we have

$$
\begin{aligned}
\widetilde{\lambda}_{\max}(k, m) &= \text{tr}\left[\phi_X(k, m)\Phi_{\text{in}}^{-1}(k, m)\Gamma_{\widetilde{\mathbf{x}}}(k, m)\mathbf{i}_1\mathbf{i}_1^T\Gamma_{\widetilde{\mathbf{x}}}(k, m)\right] \\
&= \phi_X(k, m)\mathbf{i}_1^T\Gamma_{\widetilde{\mathbf{x}}}(k, m)\Phi_{\text{in}}^{-1}(k, m)\Gamma_{\widetilde{\mathbf{x}}}(k, m)\mathbf{i}_1,
\end{aligned}
\tag{6.21}
$$

where tr[·] denotes the trace of a square matrix. As a result,

$$\text{oSNR}\left[\widetilde{\mathbf{h}}_{\max}(k, m)\right] = \widetilde{\lambda}_{\max}(k, m), \ k = 0, 1, \ldots, K - 1,
\tag{6.22}$$

which corresponds to the maximum possible output SNR according to the inequality in (6.5). Obviously, we also have

$$\widetilde{\mathbf{h}}_{\max}(k, m) = \alpha(k, m)\phi_X(k, m)\Phi_{\text{in}}^{-1}(k, m)\Gamma_{\widetilde{\mathbf{x}}}(k, m)\mathbf{i}_1, \ k = 0, 1, \ldots, K - 1,
\tag{6.23}$$

where $\alpha(k, m)$ is an arbitrary scaling factor different from zero. While this factor has no effect on the narrowband output SNR, it has on the fullband output SNR and speech distortion (narrowband and fullband).

Let us see now what happens if the desired and noise signals are circular [i.e., $\gamma_X(k, m) = \gamma_V(k, m) = 0$]. In this scenario, it is easy to verify that

$$
\begin{aligned}
\Gamma_{\widetilde{\mathbf{x}}}(k, m)\mathbf{i}_1 &= \mathbf{i}_1, \tag{6.24}\\
\Phi_{\text{in}}(k, m) &= \begin{bmatrix} \phi_V(k, m) & 0 \\ 0 & \phi_Y(k, m) \end{bmatrix}, \tag{6.25}
\end{aligned}
$$

and, as a result,

$$\widetilde{\mathbf{h}}_{\max}(k, m) = \alpha(k, m)\begin{bmatrix} \text{iSNR}(k, m) \\ 0 \end{bmatrix}.
\tag{6.26}$$

As expected, the maximum SNR filter simplifies to a gain at each frequency-bin which cannot improve the narrowband output SNR.

6.3   WIENER FILTER

Taking the gradient of the frequency-domain MSE, $J\left[\widetilde{\mathbf{h}}(k,m)\right]$, with respect to $\widetilde{\mathbf{h}}^{H}(k,m)$ and equating the result to zero gives us the Wiener filter (6):

$$
\begin{aligned}
\widetilde{\mathbf{h}}_{W}(k,m) &= \Phi_{\widetilde{\mathbf{y}}}^{-1}(k,m)\Phi_{\widetilde{\mathbf{x}}}(k,m)\mathbf{i}_1 \\
&= \phi_X(k,m)\Phi_{\widetilde{\mathbf{y}}}^{-1}(k,m)\Gamma_{\widetilde{\mathbf{x}}}(k,m)\mathbf{i}_1 \\
&= \frac{\phi_X(k,m)}{\phi_Y(k,m)} \cdot \Gamma_{\widetilde{\mathbf{y}}}^{-1}(k,m)\Gamma_{\widetilde{\mathbf{x}}}(k,m)\mathbf{i}_1 \\
&= \left[\mathbf{I}_2 - \frac{\phi_V(k,m)}{\phi_Y(k,m)} \cdot \Gamma_{\widetilde{\mathbf{y}}}^{-1}(k,m)\Gamma_{\widetilde{\mathbf{v}}}(k,m)\right]\mathbf{i}_1.
\end{aligned}
\tag{6.27}
$$

It follows immediately that the two components of the vector $\widetilde{\mathbf{h}}_{W}(k,m)$ are

$$
\widetilde{H}_{W}(k,m) = \frac{1-\gamma_X(k,m)\gamma_Y^{*}(k,m)}{1-|\gamma_Y(k,m)|^2} \cdot \frac{\phi_X(k,m)}{\phi_Y(k,m)},
\tag{6.28}
$$

$$
\widetilde{H}_{W}'(k,m) = \frac{\gamma_X(k,m)-\gamma_Y(k,m)}{1-|\gamma_Y(k,m)|^2} \cdot \frac{\phi_X(k,m)}{\phi_Y(k,m)}.
\tag{6.29}
$$

It is easy to deduce the following relation:

$$
\gamma_Y(k,m)\phi_Y(k,m) = \gamma_X(k,m)\phi_X(k,m) + \gamma_V(k,m)\phi_V(k,m).
\tag{6.30}
$$

By using (6.30), the Wiener complex gains in (6.28)–(6.29), can be rearranged as

$$
\widetilde{H}_{W}(k,m) = 1 - \frac{1-\gamma_V(k,m)\gamma_Y^{*}(k,m)}{1-|\gamma_Y(k,m)|^2} \cdot \frac{\phi_V(k,m)}{\phi_Y(k,m)},
\tag{6.31}
$$

$$
\widetilde{H}_{W}'(k,m) = \frac{\gamma_Y(k,m)-\gamma_V(k,m)}{1-|\gamma_Y(k,m)|^2} \cdot \frac{\phi_V(k,m)}{\phi_Y(k,m)}.
\tag{6.32}
$$

We recall that the classical Wiener gain is (see Chapter 5)

$$
H_{W}(k,m) = \frac{\phi_X(k,m)}{\phi_Y(k,m)} = 1 - \frac{\phi_V(k,m)}{\phi_Y(k,m)}.
\tag{6.33}
$$

Of course, taking $\gamma_X(k,m) = \gamma_V(k,m) = 0$ in the Wiener filter, we obtain the classical Wiener gain. While the conventional Wiener gain is always real, the two components of the Wiener filter obtained with Model 2 are, in general, complex.

In practice, $\gamma_X(k,m)$ is in general different from 0 since speech is inherently nonstationary. But it is not uncommon that noise is relatively stationary, and, presumably, we can take $\gamma_V(k,m) = 0$. In this case, we see from (6.31) and (6.32) that $\widetilde{H}_{W}(k,m)$ is real, but $\widetilde{H}_{W}'(k,m)$ is still complex. To study the deviation of these two complex Wiener gains from the conventional Wiener gain, we rewrite them as follows:

$$\widetilde{H}_W(k, m) = \eta_W(k, m) H_W(k, m), \tag{6.34}$$

$$\left|\widetilde{H}'_W(k, m)\right| = \frac{\dfrac{\text{iSNR}(k, m)}{[1 + \text{iSNR}(k, m)]^2} \cdot |\gamma_X(k, m)|}{1 - \left[\dfrac{\text{iSNR}(k, m)}{1 + \text{iSNR}(k, m)}\right]^2 \cdot |\gamma_X(k, m)|^2}, \tag{6.35}$$

where

$$\eta_W(k, m) = \frac{1 - \dfrac{\text{iSNR}(k, m)}{1 + \text{iSNR}(k, m)} \cdot |\gamma_X(k, m)|^2}{1 - \left[\dfrac{\text{iSNR}(k, m)}{1 + \text{iSNR}(k, m)}\right]^2 \cdot |\gamma_X(k, m)|^2}. \tag{6.36}$$

It can easily be verified that $0 \leq \eta_W(k, m) \leq 1$ and $0 \leq \left|\widetilde{H}'_W(k, m)\right| \leq 1$. These two quantities are presented in Fig. 6.1 as a function of the narrowband input SNR, $\text{iSNR}(k, m)$, and the magnitude of speech circularity quotient $|\gamma_X(k, m)|$. In the plots, we can mark out two characteristic ranges.

1. Very large $\text{iSNR}(k, m)$ (roughly ≥ 30 dB) or very small $\text{iSNR}(k, m)$ (roughly ≤ -20 dB): in this range, $\eta_W(k, m)$ approaches 1, and, meanwhile, $\left|\widetilde{H}'_W(k, m)\right|$ is close to 0. This indicates that the Model-2 Wiener filter converges to the conventional Wiener gain when either speech or noise is sufficiently dominant in the microphone output regardless of the degree of speech noncircularity.

2. Moderate $\text{iSNR}(k, m)$ (somewhere between -20 dB and 30 dB): this is the range where speech noncircularity may be useful for noise reduction (for stationary noise), and the new Model-2 Wiener filter is different from its conventional counterpart.

Let us compare the speech distortion index of the Wiener filters for Model 2 [again assuming $\gamma_V(k, m) = 0$] against that of the conventional Wiener filter. Substituting (6.34) and (6.35) into (6.13) produces

$$\upsilon_{\text{sd}}\left[\widetilde{\mathbf{h}}_W(k, m)\right] = \left|1 - \widetilde{G}_W(k, m)\right|^2, \tag{6.37}$$

where

$$\begin{aligned}
\widetilde{G}_W(k, m) &= \widetilde{H}_W(k, m) + \gamma_X^*(k, m) \widetilde{H}'_W(k, m) \\
&= \frac{1 + \dfrac{1 - \text{iSNR}(k, m)}{1 + \text{iSNR}(k, m)} \cdot |\gamma_X(k, m)|^2}{1 - \left[\dfrac{\text{iSNR}(k, m)}{1 + \text{iSNR}(k, m)}\right]^2 \cdot |\gamma_X(k, m)|^2} \cdot \frac{\text{iSNR}(k, m)}{1 + \text{iSNR}(k, m)}.
\end{aligned} \tag{6.38}$$

Recall that for the classical Wiener gain, the speech distortion index is [see eq. (5.18)]

$$\upsilon_{\text{sd}}\left[H_W(k, m)\right] = |1 - G_W(k, m)|^2, \tag{6.39}$$

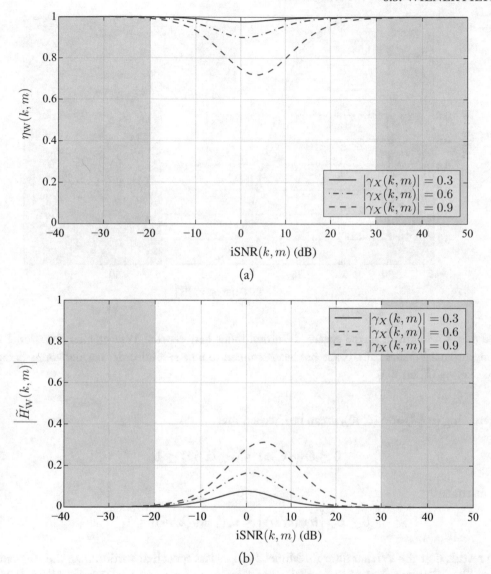

Figure 6.1: Deviation of the complex Wiener gains from the conventional Wiener gain as a function of the input SNR: (a) $\eta_W(k, m)$ and (b) $\left|\widetilde{H}'_W(k, m)\right|$. Noise has been assumed to be a second-order circular complex random variable, i.e., $\gamma_V(k, m) = 0$.

where

$$
\begin{aligned}
G_W(k, m) &= H_W(k, m) \\
&= \frac{\text{iSNR}(k, m)}{1 + \text{iSNR}(k, m)}.
\end{aligned}
\tag{6.40}
$$

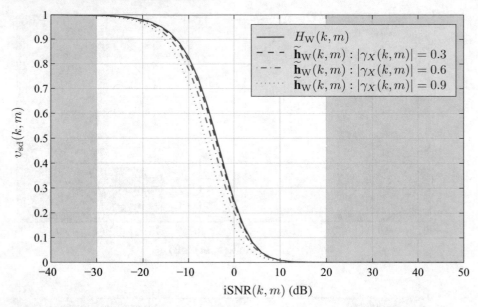

Figure 6.2: Comparison in the speech distortion index between the Wiener filter for Model 2 and the conventional Wiener gain. Noise has been assumed to be a second-order circular complex random variable, i.e., $\gamma_V(k, m) = 0$.

By examining (6.38) and (6.40), it can be checked that

$$0 \leq G_{\mathrm{W}}(k, m) \leq \widetilde{G}_{\mathrm{W}}(k, m) \leq 1, \tag{6.41}$$

and consequently

$$0 \leq \upsilon_{\mathrm{sd}} \left[\widetilde{\mathbf{h}}_{\mathrm{W}}(k, m) \right] \leq \upsilon_{\mathrm{sd}} \left[H_{\mathrm{W}}(k, m) \right] \leq 1, \tag{6.42}$$

which reveals that the Wiener filter for Model 2 causes less speech distortion than the conventional Wiener filter. Figure 6.2 plots the speech distortion index for these two Wiener filters. It is clear that speech distortion of these two Wiener filters decreases monotonically as the narrowband input SNR increases. It can also be checked from (6.37) and (6.38) that for a fixed narrowband input SNR, a larger circularity quotient $\gamma_X(k, m)$ leads to a small speech distortion index. But when the input SNR is either very large or very small, the improvement in speech distortion index due to the use of speech noncircularity is insignificant.

There is another interesting way to write the Wiener filter. Indeed, it is easy to verify that

$$\Phi_{\widetilde{\mathbf{y}}}(k, m) = \phi_X(k, m) \Gamma_{\widetilde{\mathbf{x}}}(k, m) \mathbf{i}_1 \mathbf{i}_1^T \Gamma_{\widetilde{\mathbf{x}}}(k, m) + \Phi_{\mathrm{in}}(k, m). \tag{6.43}$$

Determining the inverse of $\Phi_{\widetilde{\mathbf{y}}}(k, m)$ from (6.43) with the Woodbury's identity

$$\Phi_{\widetilde{\mathbf{y}}}^{-1}(k, m) = \Phi_{\text{in}}^{-1}(k, m) - \frac{\Phi_{\text{in}}^{-1}(k, m)\Gamma_{\widetilde{\mathbf{x}}}(k, m)\mathbf{i}_1\mathbf{i}_1^T\Gamma_{\widetilde{\mathbf{x}}}(k, m)\Phi_{\text{in}}^{-1}(k, m)}{\phi_X^{-1}(k, m) + \mathbf{i}_1^T\Gamma_{\widetilde{\mathbf{x}}}(k, m)\Phi_{\text{in}}^{-1}(k, m)\Gamma_{\widetilde{\mathbf{x}}}(k, m)\mathbf{i}_1} \tag{6.44}$$

and substituting the result into (6.27), leads to another formulation of the Wiener filter:

$$\begin{aligned}
\widetilde{\mathbf{h}}_W(k, m) &= \frac{\Phi_{\text{in}}^{-1}(k, m)\Gamma_{\widetilde{\mathbf{x}}}(k, m)\mathbf{i}_1}{\phi_X^{-1}(k, m) + \mathbf{i}_1^T\Gamma_{\widetilde{\mathbf{x}}}(k, m)\Phi_{\text{in}}^{-1}(k, m)\Gamma_{\widetilde{\mathbf{x}}}(k, m)\mathbf{i}_1} \\
&= \frac{\phi_X(k, m)\Phi_{\text{in}}^{-1}(k, m)\Gamma_{\widetilde{\mathbf{x}}}(k, m)\mathbf{i}_1}{1 + \widetilde{\lambda}_{\max}(k, m)}.
\end{aligned} \tag{6.45}$$

We can deduce from (6.45) that the narrowband output SNR is

$$\text{oSNR}\left[\widetilde{\mathbf{h}}_W(k, m)\right] = \widetilde{\lambda}_{\max}(k, m), \ k = 0, 1, \ldots, K - 1 \tag{6.46}$$

and the narrowband speech distortion index is a clear function of the narrowband output SNR:

$$\upsilon_{\text{sd}}\left[\widetilde{\mathbf{h}}_W(k, m)\right] = \frac{1}{\left\{1 + \text{oSNR}\left[\widetilde{\mathbf{h}}_W(k, m)\right]\right\}^2}. \tag{6.47}$$

The higher is the value of $\text{oSNR}\left[\widetilde{\mathbf{h}}_W(k, m)\right]$, the less the desired signal is distorted.

Property 6.1   With the frequency-domain Wiener filter given in (6.27), the narrowband output SNR is always greater than or equal to the narrowband input SNR, i.e., $\text{oSNR}\left[\widetilde{\mathbf{h}}_W(k, m)\right] \geq \text{iSNR}(k, m)$.

Proof. This result is straightforward since the Wiener filter maximizes the narrowband output SNR.
□

Recall that in Model 1, the narrowband output SNR cannot be improved.

It is of great interest to observe that the two filters, $\widetilde{\mathbf{h}}_{\max}(k, m)$ and $\widetilde{\mathbf{h}}_W(k, m)$, are equivalent up to a scaling factor. Indeed, taking

$$\alpha(k, m) = \frac{1}{1 + \widetilde{\lambda}_{\max}(k, m)} \tag{6.48}$$

in (6.23) (maximum SNR filter), we find (6.45) (Wiener filter).

With the Wiener filter, the narrowband noise reduction factor is

$$\xi_{\mathrm{nr}}\left[\widetilde{\mathbf{h}}_{\mathrm{W}}(k,m)\right] = \frac{\left\{1 + \mathrm{oSNR}\left[\widetilde{\mathbf{h}}_{\mathrm{W}}(k,m)\right]\right\}^2}{\mathrm{iSNR}(k,m)\cdot\mathrm{oSNR}\left[\widetilde{\mathbf{h}}_{\mathrm{W}}(k,m)\right]} \tag{6.49}$$

$$\geq \left\{1 + \frac{1}{\mathrm{oSNR}\left[\widetilde{\mathbf{h}}_{\mathrm{W}}(k,m)\right]}\right\}^2.$$

Using (6.47) and (6.49) in the NMSE, we find the minimum NMSE:

$$\widetilde{J}\left[\widetilde{\mathbf{h}}_{\mathrm{W}}(k,m)\right] = \frac{\mathrm{iSNR}(k,m)}{1 + \mathrm{oSNR}\left[\widetilde{\mathbf{h}}_{\mathrm{W}}(k,m)\right]} \leq 1. \tag{6.50}$$

The fullband output SNR is

$$\mathrm{oSNR}\left[\widetilde{\mathbf{h}}_{\mathrm{W}}(:,m)\right] = \frac{\sum_{k=0}^{K-1}\phi_X(k,m)\dfrac{\mathrm{oSNR}^2\left[\widetilde{\mathbf{h}}_{\mathrm{W}}(k,m)\right]}{\left\{1 + \mathrm{oSNR}\left[\widetilde{\mathbf{h}}_{\mathrm{W}}(k,m)\right]\right\}^2}}{\sum_{k=0}^{K-1}\phi_X(k,m)\dfrac{\mathrm{oSNR}\left[\widetilde{\mathbf{h}}_{\mathrm{W}}(k,m)\right]}{\left\{1 + \mathrm{oSNR}\left[\widetilde{\mathbf{h}}_{\mathrm{W}}(k,m)\right]\right\}^2}}. \tag{6.51}$$

Property 6.2 With the frequency-domain Wiener filter given in (6.27), the fullband output SNR is always greater than or equal to the fullband input SNR, i.e., $\mathrm{oSNR}\left[\widetilde{\mathbf{h}}_{\mathrm{W}}(:,m)\right] \geq \mathrm{iSNR}(m)$.

Proof. See Section 6.5. □

6.4 MINIMUM VARIANCE DISTORTIONLESS RESPONSE (MVDR) FILTER

The celebrated minimum variance distortionless response (MVDR) filter, proposed by Capon (13), (29) is usually derived in a context where we have at least two sensors (or microphones) available. Interestingly, with Model 2, we can also derive the MVDR (with one sensor only) by minimizing the MSE of the residual interference-plus-noise, $J_{\mathrm{r}}\left[\widetilde{\mathbf{h}}(k,m)\right]$, with the constraint that the desired signal is not distorted (7). The derivation of such an MVDR is possible, thanks to the noncircularity of the signals. Mathematically, this is equivalent to

$$\min_{\widetilde{\mathbf{h}}(k,m)} \widetilde{\mathbf{h}}^H(k,m)\Phi_{\mathrm{in}}(k,m)\widetilde{\mathbf{h}}(k,m) \quad \text{subject to} \quad \widetilde{\mathbf{h}}^H(k,m)\Gamma_{\widetilde{\mathbf{x}}}(k,m)\mathbf{i}_1 = 1, \tag{6.52}$$

for which the solution is

$$\widetilde{\mathbf{h}}_{\text{MVDR}}(k, m) = \frac{\phi_X(k, m)\Phi_{\text{in}}^{-1}(k, m)\Gamma_{\widetilde{\mathbf{x}}}(k, m)\mathbf{i}_1}{\widetilde{\lambda}_{\max}(k, m)}.$$ (6.53)

Obviously, we can rewrite the MVDR as

$$\widetilde{\mathbf{h}}_{\text{MVDR}}(k, m) = \frac{\Phi_{\widetilde{\mathbf{y}}}^{-1}(k, m)\Gamma_{\widetilde{\mathbf{x}}}(k, m)\mathbf{i}_1}{\mathbf{i}_1^T \Gamma_{\widetilde{\mathbf{x}}}(k, m)\Phi_{\widetilde{\mathbf{y}}}^{-1}(k, m)\Gamma_{\widetilde{\mathbf{x}}}(k, m)\mathbf{i}_1}.$$ (6.54)

If the signals are circular [i.e., $\gamma_X(k, m) = \gamma_V(k, m) = 0$], then $\widetilde{\mathbf{h}}_{\text{MVDR}}(k, m) = \mathbf{i}_1$; the MVDR filter simplifies to the identity filter, and it neither improves the output SNR nor distorts the speech signal.

Taking

$$\alpha(k, m) = \frac{1}{\widetilde{\lambda}_{\max}(k, m)}$$ (6.55)

in (6.23) (maximum SNR filter), we find (6.53) (MVDR filter), showing how the maximum SNR, MVDR, and Wiener filters are equivalent up to a scaling factor. From a narrowband point of view, this scaling is not significant, but from a fullband point of view, it can be important since speech signals are broadband in nature. Indeed, it can easily be verified that this scaling factor affects the fullband output SNRs and fullband speech distortion indices. While the narrowband output SNRs of the maximum SNR, Wiener, and MVDR filters are the same, the fullband output SNRs are not because of the scaling factor.

It is clear that we always have

$$\text{oSNR}\left[\widetilde{\mathbf{h}}_{\text{MVDR}}(k, m)\right] = \text{oSNR}\left[\widetilde{\mathbf{h}}_{\text{W}}(k, m)\right],$$ (6.56)

$$\upsilon_{\text{sd}}\left[\widetilde{\mathbf{h}}_{\text{MVDR}}(k, m)\right] = 0,$$ (6.57)

$$\xi_{\text{sr}}\left[\widetilde{\mathbf{h}}_{\text{MVDR}}(k, m)\right] = 1,$$ (6.58)

$$\xi_{\text{nr}}\left[\widetilde{\mathbf{h}}_{\text{MVDR}}(k, m)\right] = \frac{\widetilde{\lambda}_{\max}(k, m)}{\text{iSNR}(k, m)} \leq \xi_{\text{nr}}\left[\widetilde{\mathbf{h}}_{\text{W}}(k, m)\right],$$ (6.59)

and

$$1 \geq \widetilde{J}\left[\widetilde{\mathbf{h}}_{\text{MVDR}}(k, m)\right] = \frac{\text{iSNR}(k, m)}{\widetilde{\lambda}_{\max}(k, m)} \geq \frac{\text{iSNR}(k, m)}{1 + \widetilde{\lambda}_{\max}(k, m)} = \widetilde{J}\left[\widetilde{\mathbf{h}}_{\text{W}}(k, m)\right].$$ (6.60)

We easily find that the narrowband SNR gain is

$$\begin{aligned}\mathcal{A}(k, m) &= \frac{\text{oSNR}\left[\widetilde{\mathbf{h}}_{\text{MVDR}}(k, m)\right]}{\text{iSNR}(k, m)} \\ &= 1 + \frac{|\gamma_X(k, m) - \gamma_V(k, m)|^2}{1 - |\gamma_V(k, m)|^2 + \text{iSNR}(k, m)\left[1 - |\gamma_X(k, m)|^2\right]} \geq 1,\end{aligned}$$ (6.61)

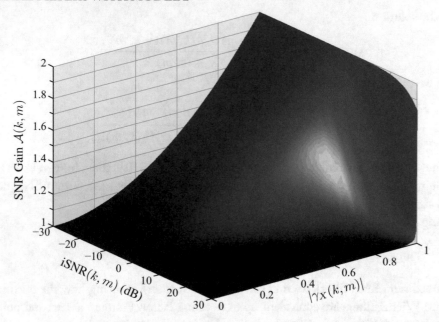

Figure 6.3: Theoretical gain in the narrowband SNR, $\mathcal{A}(k, m)$, of the MVDR filter for Model 2 as a function of the narrowband input SNR, iSNR(k, m), and the magnitude of the speech circularity quotient, $|\gamma_X(k, m)|$. Noise has been assumed to be a second-order circular complex random variable, i.e., $\gamma_V(k, m) = 0$.

with equality if and only if $\gamma_X(k, m) = \gamma_V(k, m) = 0$ (i.e., circular signals).

In a more practical situation where only the noise is assumed stationary [i.e., $\gamma_V(k, m) = 0$ but $\gamma_X(k, m) \neq 0$], we obtain from (6.61) that

$$1 \leq \mathcal{A}(k, m) = 1 + \frac{|\gamma_X(k, m)|^2}{1 + \text{iSNR}(k, m)\left[1 - |\gamma_X(k, m)|^2\right]} \leq 2. \qquad (6.62)$$

The narrowband SNR gain by the MVDR filter is upper bounded by 2 (approximately 3 dB). Figure 6.3 visualizes the relationship of this gain with iSNR(k, m) and $|\gamma_X(k, m)|$. When $|\gamma_X(k, m)|$ is small, speech circularity is too trivial to be exploited, and the narrowband SNR cannot be improved by the MVDR filter, which is similar to the conventional techniques for Model 1 (4). On the other hand, when the input SNR is very high, the improvement due to the use of speech noncircularity is also marginal. The narrowband SNR gain by the MVDR filter is significant when the input SNR is low and the magnitude of the speech circularity quotient is large.

The fullband output SNR is

$$\text{oSNR}\left[\widetilde{\mathbf{h}}_{\text{MVDR}}(:,m)\right] = \frac{\sum_{k=0}^{K-1}\phi_X(k,m)}{\sum_{k=0}^{K-1}\dfrac{\phi_X(k,m)}{\text{oSNR}\left[\widetilde{\mathbf{h}}_{\text{MVDR}}(:,m)\right]}}. \tag{6.63}$$

Property 6.3 With the optimal frequency-domain MVDR filter given in (6.53), the fullband output SNR is always greater than or equal to the fullband input SNR, i.e., $\text{oSNR}\left[\widetilde{\mathbf{h}}_{\text{MVDR}}(:,m)\right] \geq \text{iSNR}(m)$.

Proof. See Section 6.5. □

6.5   TRADEOFF FILTER

As we have learned from the two previous sections, not much flexibility is associated with the Wiener and MVDR filters in the sense that we do not know in advance by how much the output SNR will be improved. However, in many practical situations, we wish to have some flexibility to control the compromise between noise reduction and speech distortion, and the best way to do this is via the so-called tradeoff filter.

The narrowband MSE is the sum of two terms. One depends on the speech distortion while the other one depends on the residual interference-plus-noise. Instead of minimizing the narrowband MSE as we already did to find the Wiener filter, we can minimize the MSE due to the speech distortion with the constraint that the noise reduction factor is equal to a positive value that is greater than one. Mathematically, this is equivalent to

$$\min_{\widetilde{\mathbf{h}}(k,m)} J_d\left[\widetilde{\mathbf{h}}(k,m)\right] \quad \text{subject to} \quad J_r\left[\widetilde{\mathbf{h}}(k,m)\right] = \beta\phi_V(k,m), \tag{6.64}$$

where $0 < \beta < 1$ to insure that we get some noise reduction. By using a Lagrange multiplier, $\mu \geq 0$, to adjoin the constraint to the cost function, we easily deduce the tradeoff filter (6):

$$\begin{aligned}\widetilde{\mathbf{h}}_{\text{T},\mu}(k,m) &= \phi_X(k,m)\left[\phi_X(k,m)\Gamma_{\widetilde{\mathbf{x}}}(k,m)\mathbf{i}_1\mathbf{i}_1^T\Gamma_{\widetilde{\mathbf{x}}}(k,m) + \mu\Phi_{\text{in}}(k,m)\right]^{-1}\Gamma_{\widetilde{\mathbf{x}}}(k,m)\mathbf{i}_1 \\ &= \frac{\phi_X(k,m)\Phi_{\text{in}}^{-1}(k,m)\Gamma_{\widetilde{\mathbf{x}}}(k,m)\mathbf{i}_1}{\mu + \widetilde{\lambda}_{\max}(k,m)}, \end{aligned} \tag{6.65}$$

where the Lagrange multiplier, μ, satisfies $J_r\left[\widetilde{\mathbf{h}}_{\text{T},\mu}(k,m)\right] = \beta\phi_V(k,m)$. However, in practice, it is not easy to determine the optimal μ. Therefore, when this parameter is chosen in an ad-hoc way, we can see that for

- $\mu = 1$, $\widetilde{\mathbf{h}}_{\text{T},1}(k,m) = \widetilde{\mathbf{h}}_{\text{W}}(k,m)$, which is the Wiener filter;

- $\mu = 0$, $\widetilde{\mathbf{h}}_{T,0}(k, m) = \widetilde{\mathbf{h}}_{MVDR}(k, m)$, which is the MVDR filter;

- $\mu > 1$, results in a filter with low residual noise at the expense of high speech distortion;

- $\mu < 1$, results in a filter with high residual noise and low speech distortion.

Again, we observe here as well that the tradeoff and Wiener filters are equivalent up to a scaling factor. As a result, the narrowband output SNR with the tradeoff filter is obviously the same as the narrowband output SNR with the Wiener filter, i.e.,

$$\mathrm{oSNR}\left[\widetilde{\mathbf{h}}_{T,\mu}(k, m)\right] = \widetilde{\lambda}_{\max}(k, m), \tag{6.66}$$

and does not depend on μ. However, the narrowband speech distortion index is now both a function of the variable μ and the narrowband output SNR:

$$\upsilon_{sd}\left[\widetilde{\mathbf{h}}_{T,\mu}(k, m)\right] = \frac{\mu^2}{\left[\mu + \widetilde{\lambda}_{\max}(k, m)\right]^2}. \tag{6.67}$$

From (6.67), we observe how μ can affect the desired signal.

The tradeoff filter is interesting from several perspectives since it encompasses both the Wiener and MVDR filters. It is then useful to study the fullband output SNR and the fullband speech distortion index of the tradeoff filter, which both depend on the variable μ.

Using (6.65) in (6.7), we find that the fullband output SNR is

$$\mathrm{oSNR}\left[\widetilde{\mathbf{h}}_{T,\mu}(:, m)\right] = \frac{\displaystyle\sum_{k=0}^{K-1} \frac{\phi_X(k, m)\widetilde{\lambda}_{\max}^2(k, m)}{\left[\mu + \widetilde{\lambda}_{\max}(k, m)\right]^2}}{\displaystyle\sum_{k=0}^{K-1} \frac{\phi_X(k, m)\widetilde{\lambda}_{\max}(k, m)}{\left[\mu + \widetilde{\lambda}_{\max}(k, m)\right]^2}}. \tag{6.68}$$

We propose the following.

Property 6.4   The fullband output SNR of the tradeoff filter is an increasing function of the parameter μ.

Proof. The proof presented here is very similar to the one given in (43).

In order to determine the variations of $\mathrm{oSNR}\left[\widetilde{\mathbf{h}}_{T,\mu}(:, m)\right]$ with respect to the parameter μ, we will check the sign of the following differentiation with respect to μ:

$$\frac{d\mathrm{oSNR}\left[\widetilde{\mathbf{h}}_{T,\mu}(:, m)\right]}{d\mu} = 2\frac{\mathrm{Num}(\mu)}{\mathrm{Den}(\mu)}, \tag{6.69}$$

where

$$
\begin{aligned}
\text{Num}(\mu) &= -\sum_{k=0}^{K-1} \frac{\phi_X(k,m)\widetilde{\lambda}_{\max}(k,m)}{\left[\mu+\widetilde{\lambda}_{\max}(k,m)\right]^2} \sum_{k=0}^{K-1} \frac{\phi_X(k,m)\widetilde{\lambda}_{\max}^2(k,m)}{\left[\mu+\widetilde{\lambda}_{\max}(k,m)\right]^3} \\
&\quad + \sum_{k=0}^{K-1} \frac{\phi_X(k,m)\widetilde{\lambda}_{\max}^2(k,m)}{\left[\mu+\widetilde{\lambda}_{\max}(k,m)\right]^2} \sum_{k=0}^{K-1} \frac{\phi_X(k,m)\widetilde{\lambda}_{\max}(k,m)}{\left[\mu+\widetilde{\lambda}_{\max}(k,m)\right]^3}, \qquad (6.70)
\end{aligned}
$$

$$
\text{Den}(\mu) = \left\{ \sum_{k=0}^{K-1} \frac{\phi_X(k,m)\widetilde{\lambda}_{\max}(k,m)}{\left[\mu+\widetilde{\lambda}_{\max}(k,m)\right]^2} \right\}^2. \qquad (6.71)
$$

We only focus on the numerator of the above derivative to see the variations of the fullband output SNR since the denominator is always positive. Multiplying and dividing by $\mu+\widetilde{\lambda}_{\max}(k,m)$, this numerator can be rewritten as

$$
\begin{aligned}
\text{Num}(\mu) &= -\sum_{k=0}^{K-1} \frac{\phi_X(k,m)\widetilde{\lambda}_{\max}(k,m)\left[\mu+\widetilde{\lambda}_{\max}(k,m)\right]}{\left[\mu+\widetilde{\lambda}_{\max}(k,m)\right]^3} \sum_{k=0}^{K-1} \frac{\phi_X(k,m)\widetilde{\lambda}_{\max}^2(k,m)}{\left[\mu+\widetilde{\lambda}_{\max}(k,m)\right]^3} \\
&\quad + \sum_{k=0}^{K-1} \frac{\phi_X(k,m)\widetilde{\lambda}_{\max}^2(k,m)\left[\mu+\widetilde{\lambda}_{\max}(k,m)\right]}{\left[\mu+\widetilde{\lambda}_{\max}(k,m)\right]^3} \sum_{k=0}^{K-1} \frac{\phi_X(k,m)\widetilde{\lambda}_{\max}(k,m)}{\left[\mu+\widetilde{\lambda}_{\max}(k,m)\right]^3} \\
&= -\left\{ \sum_{k=0}^{K-1} \frac{\phi_X(k,m)\widetilde{\lambda}_{\max}^2(k,m)}{\left[\mu+\widetilde{\lambda}_{\max}(k,m)\right]^3} \right\}^2 \\
&\quad -\mu \sum_{k=0}^{K-1} \frac{\phi_X(k,m)\widetilde{\lambda}_{\max}(k,m)}{\left[\mu+\widetilde{\lambda}_{\max}(k,m)\right]^3} \sum_{k=0}^{K-1} \frac{\phi_X(k,m)\widetilde{\lambda}_{\max}^2(k,m)}{\left[\mu+\widetilde{\lambda}_{\max}(k,m)\right]^3} \\
&\quad + \sum_{k=0}^{K-1} \frac{\phi_X(k,m)\widetilde{\lambda}_{\max}^3(k,m)}{\left[\mu+\widetilde{\lambda}_{\max}(k,m)\right]^3} \sum_{k=0}^{K-1} \frac{\phi_X(k,m)\widetilde{\lambda}_{\max}(k,m)}{\left[\mu+\widetilde{\lambda}_{\max}(k,m)\right]^3} \\
&\quad +\mu \sum_{k=0}^{K-1} \frac{\phi_X(k,m)\widetilde{\lambda}_{\max}(k,m)}{\left[\mu+\widetilde{\lambda}_{\max}(k,m)\right]^3} \sum_{k=0}^{K-1} \frac{\phi_X(k,m)\widetilde{\lambda}_{\max}^2(k,m)}{\left[\mu+\widetilde{\lambda}_{\max}(k,m)\right]^3} \\
&= -\left\{ \sum_{k=0}^{K-1} \frac{\phi_X(k,m)\widetilde{\lambda}_{\max}^2(k,m)}{\left[\mu+\widetilde{\lambda}_{\max}(k,m)\right]^3} \right\}^2 \\
&\quad + \sum_{k=0}^{K-1} \frac{\phi_X(k,m)\widetilde{\lambda}_{\max}^3(k,m)}{\left[\mu+\widetilde{\lambda}_{\max}(k,m)\right]^3} \sum_{k=0}^{K-1} \frac{\phi_X(k,m)\widetilde{\lambda}_{\max}(k,m)}{\left[\mu+\widetilde{\lambda}_{\max}(k,m)\right]^3}. \qquad (6.72)
\end{aligned}
$$

As far as $\mu, \widetilde{\lambda}_{\max}(k, m)$, and $\phi_X(k, m)$ are positive $\forall k, m$, we can use the Cauchy-Schwarz inequality

$$
\begin{aligned}
\sum_{k=0}^{K-1} & \frac{\phi_X(k, m)\widetilde{\lambda}_{\max}^3(k, m)}{\left[\mu + \widetilde{\lambda}_{\max}(k, m)\right]^3} \sum_{k=0}^{K-1} \frac{\phi_X(k, m)\widetilde{\lambda}_{\max}(k, m)}{\left[\mu + \widetilde{\lambda}_{\max}(k, m)\right]^3} \\
&\geq \left\{\sum_{k=0}^{K-1} \sqrt{\frac{\phi_X(k, m)\widetilde{\lambda}_{\max}^3(k, m)}{\left[\mu + \widetilde{\lambda}_{\max}(k, m)\right]^3}} \sqrt{\frac{\phi_X(k, m)\widetilde{\lambda}_{\max}(k, m)}{\left[\mu + \widetilde{\lambda}_{\max}(k, m)\right]^3}}\right\}^2 \\
&= \left\{\sum_{k=0}^{K-1} \frac{\phi_X(k, m)\widetilde{\lambda}_{\max}^2(k, m)}{\left[\mu + \widetilde{\lambda}_{\max}(k, m)\right]^3}\right\}^2.
\end{aligned}
\tag{6.73}
$$

Substituting (6.73) into (6.72), we conclude that

$$
\frac{d\text{oSNR}\left[\widetilde{\mathbf{h}}_{\text{T},\mu}(:, m)\right]}{d\mu} \geq 0,
\tag{6.74}
$$

proving that the fullband output SNR is increasing with respect to μ. $\square$

From Property 6.4, we deduce that the MVDR filter gives the smallest fullband output SNR, which is

$$
\text{oSNR}\left[\widetilde{\mathbf{h}}_{\text{T},0}(:, m)\right] = \frac{\sum_{k=0}^{K-1} \phi_X(k, m)}{\sum_{k=0}^{K-1} \dfrac{\phi_X(k, m)}{\widetilde{\lambda}_{\max}(k, m)}}.
\tag{6.75}
$$

We give another interesting property.

Property 6.5   We have

$$
\lim_{\mu \to \infty} \text{oSNR}\left[\widetilde{\mathbf{h}}_{\text{T},\mu}(:, m)\right] = \frac{\sum_{k=0}^{K-1} \phi_X(k, m)\widetilde{\lambda}_{\max}^2(k, m)}{\sum_{k=0}^{K-1} \phi_X(k, m)\widetilde{\lambda}_{\max}(k, m)} \leq \sum_{k=0}^{K-1} \widetilde{\lambda}_{\max}(k, m).
\tag{6.76}
$$

Proof. Easy to show from (6.68). $\square$

While the fullband output SNR is upper bounded, it is easy to show that the fullband noise reduction factor and fullband speech reduction factor are not. So when μ goes to infinity so are $\xi_{\text{nr}}\left[\widetilde{\mathbf{h}}_{\text{T},\mu}(:, m)\right]$ and $\xi_{\text{sr}}\left[\widetilde{\mathbf{h}}_{\text{T},\mu}(:, m)\right]$.

The fullband speech distortion index is

$$
\upsilon_{\text{sd}}\left[\widetilde{\mathbf{h}}_{\text{T},\mu}(:, m)\right] = \frac{\sum_{k=0}^{K-1} \dfrac{\phi_X(k, m)\mu^2}{\left[\mu + \widetilde{\lambda}_{\max}(k, m)\right]^2}}{\sum_{k=0}^{K-1} \phi_X(k, m)}.
\tag{6.77}
$$

Property 6.6   The fullband speech distortion index of the tradeoff filter is an increasing function of the parameter μ.

Proof. It is straightforward to verify that

$$\frac{d\upsilon_{\mathrm{sd}}\left[\widetilde{\mathbf{h}}_{\mathrm{T},\mu}(:,m)\right]}{d\mu} \geq 0, \tag{6.78}$$

which ends the proof. $\square$

It is clear that

$$0 \leq \upsilon_{\mathrm{sd}}\left[\widetilde{\mathbf{h}}_{\mathrm{T},\mu}(:,m)\right] \leq 1, \ \forall \mu \geq 0. \tag{6.79}$$

Therefore, as μ increases, the fullband output SNR increases at the price of more distortion to the desired signal.

Property 6.7   With the tradeoff filter, $\widetilde{\mathbf{h}}_{\mathrm{T},\mu}(:,m)$, the fullband output SNR is always greater than or equal to the fullband input SNR, i.e., $\mathrm{oSNR}\left[\widetilde{\mathbf{h}}_{\mathrm{T},\mu}(:,m)\right] \geq \mathrm{iSNR}, \ \forall \mu \geq 0$.

Proof. We know that

$$\widetilde{\lambda}_{\max}(k,m) \geq \mathrm{iSNR}(k,m), \tag{6.80}$$

which implies that

$$\sum_{k=0}^{K-1} \phi_V(k,m)\frac{\mathrm{iSNR}(k,m)}{\widetilde{\lambda}_{\max}(k,m)} \leq \sum_{k=0}^{K-1} \phi_V(k,m) \tag{6.81}$$

and, hence,

$$
\begin{aligned}
\mathrm{oSNR}\left[\widetilde{\mathbf{h}}_{\mathrm{T},0}(:,m)\right] &= \frac{\sum_{k=0}^{K-1} \phi_X(k,m)}{\sum_{k=0}^{K-1} \phi_V(k,m)\dfrac{\mathrm{iSNR}(k,m)}{\widetilde{\lambda}_{\max}(k,m)}} \\
&\geq \frac{\sum_{k=0}^{K-1} \phi_X(k,m)}{\sum_{k=0}^{K-1} \phi_V(k,m)} = \mathrm{iSNR}(m).
\end{aligned}
\tag{6.82}
$$

But from Proposition 6.4, we have

$$\mathrm{oSNR}\left[\widetilde{\mathbf{h}}_{\mathrm{T},\mu}(:,m)\right] \geq \mathrm{oSNR}\left[\widetilde{\mathbf{h}}_{\mathrm{T},0}(:,m)\right], \ \forall \mu \geq 0, \tag{6.83}$$

as a result,

$$\mathrm{oSNR}\left[\widetilde{\mathbf{h}}_{\mathrm{T},\mu}(:,m)\right] \geq \mathrm{iSNR}(m), \ \forall \mu \geq 0, \tag{6.84}$$

which completes the proof. $\square$

CHAPTER 7

Optimal Filters with Model 3

In Model 3, the interframe correlation is taken into account. Thanks to this information, often underestimated in the literature, we can derive useful algorithms such as the MVDR, which is not possible with the classical method (i.e., Model 1). This new perspective is extremely important not only from a theoretical point of view but also from a practical point of view.

7.1 PERFORMANCE MEASURES

All performance measures are directly derived from Chapter 3 and Section 4.3, Chapter 4.

To quantify the level of noise remaining at the output of the FIR filter, we define the narrowband output SNR as

$$
\begin{aligned}
\text{oSNR}\left[\mathbf{h}(k,m)\right] &= \frac{\phi_{X_{\text{ld}}}(k,m)}{\phi_{X'_{\text{ri}}}(k,m) + \phi_{V_{\text{rn}}}(k,m)} \\
&= \frac{\phi_X(k,m)\left|\mathbf{h}^H(k,m)\boldsymbol{\rho}_X^*(k,m)\right|^2}{\mathbf{h}^H(k,m)\Phi_{\text{in}}(k,m)\mathbf{h}(k,m)}, \quad k = 0, 1, \ldots, K-1,
\end{aligned} \tag{7.1}
$$

where

$$
\Phi_{\text{in}}(k,m) = \Phi_{\mathbf{x}'}(k,m) + \Phi_{\mathbf{v}}(k,m) \tag{7.2}
$$

is the interference-plus-noise covariance matrix. For the particular filter $\mathbf{h}(k,m) = \mathbf{i}_{L,1}$ (identity filter), where $\mathbf{i}_{L,1}$ is the first column of the identity matrix $\mathbf{I}_L$ (of size $L \times L$), we have

$$
\text{oSNR}\left[\mathbf{i}_{L,1}(k,m)\right] = \text{iSNR}(k,m). \tag{7.3}
$$

And for the particular case $L = 1$, we also have

$$
\text{oSNR}\left[H_0(k,m)\right] = \text{iSNR}(k,m). \tag{7.4}
$$

Hence, in the two previous scenarios, the narrowband SNR cannot be improved.

Now, let us define the quantity

$$
\begin{aligned}
\text{oSNR}_{\text{max}}(k,m) &= \text{tr}\left[\Phi_{\text{in}}^{-1}(k,m)\Phi_{\mathbf{x}_{\text{d}}}(k,m)\right] \\
&= \phi_X(k,m)\boldsymbol{\rho}_X^T(k,m)\Phi_{\text{in}}^{-1}(k,m)\boldsymbol{\rho}_X^*(k,m).
\end{aligned} \tag{7.5}
$$

This quantity corresponds to the maximum eigenvalue, $\lambda_{\max}(k, m)$, of the matrix $\Phi_{\text{in}}^{-1}(k, m)\Phi_{\mathbf{x}_d}(k, m)$. It also corresponds to the maximum output SNR since the filter, $\mathbf{h}_{\max}(k, m)$, that maximizes oSNR$\left[\mathbf{h}(k, m)\right]$ [eq. (7.1)] is the maximum eigenvector of $\Phi_{\text{in}}^{-1}(k, m)\Phi_{\mathbf{x}_d}(k, m)$ for which its corresponding eigenvalue is $\lambda_{\max}(k, m)$. As a result, we have

$$\text{oSNR}\left[\mathbf{h}(k, m)\right] \leq \text{oSNR}_{\max}(k, m) = \lambda_{\max}(k, m), \ \forall \mathbf{h}(k, m) \tag{7.6}$$

and

$$\text{oSNR}_{\max}(k, m) = \text{oSNR}\left[\mathbf{h}_{\max}(k, m)\right] \geq \text{oSNR}\left[\mathbf{i}_{L,1}(k, m)\right] = \text{iSNR}(k, m). \tag{7.7}$$

We define the fullband output SNR at time-frame m as

$$\text{oSNR}\left[\mathbf{h}(:, m)\right] = \frac{\sum_{k=0}^{K-1} \phi_X(k, m) \left|\mathbf{h}^H(k, m)\boldsymbol{\rho}_X^*(k, m)\right|^2}{\sum_{k=0}^{K-1} \mathbf{h}^H(k, m)\Phi_{\text{in}}(k, m)\mathbf{h}(k, m)} \tag{7.8}$$

and it can be verified that (4)

$$\text{oSNR}\left[\mathbf{h}(:, m)\right] \leq \sum_{k=0}^{K-1} \text{oSNR}\left[\mathbf{h}(k, m)\right]. \tag{7.9}$$

As a result,

$$\text{oSNR}\left[\mathbf{h}(:, m)\right] \leq \sum_{k=0}^{K-1} \text{oSNR}_{\max}(k, m). \tag{7.10}$$

The noise reduction factor (3), (14) quantifies the amount of noise whose is rejected by the filter. The narrowband and fullband noise reduction factors are then

$$\xi_{\text{nr}}\left[\mathbf{h}(k, m)\right] = \frac{\phi_V(k, m)}{\phi_{X_{\text{ri}}'}(k, m) + \phi_{V_{\text{rn}}}(k, m)}$$

$$= \frac{\phi_V(k, m)}{\mathbf{h}^H(k, m)\Phi_{\text{in}}(k, m)\mathbf{h}(k, m)}, \ k = 0, 1, \ldots, K - 1, \tag{7.11}$$

$$\xi_{\text{nr}}\left[\mathbf{h}(:, m)\right] = \frac{\sum_{k=0}^{K-1} \phi_V(k, m)}{\sum_{k=0}^{K-1} \mathbf{h}^H(k, m)\Phi_{\text{in}}(k, m)\mathbf{h}(k, m)}. \tag{7.12}$$

The noise reduction factors are expected to be lower bounded by 1 for optimal filters. So the more the noise is reduced, the higher are the values of the noise reduction factors.

In the same manner, we define the narrowband and fullband speech reduction factors as

$$\xi_{\text{sr}}\left[\mathbf{h}(k,m)\right] = \frac{\phi_X(k,m)}{\phi_{X_{\text{ld}}}(k,m)}$$

$$= \frac{1}{\left|\mathbf{h}^H(k,m)\boldsymbol{\rho}_X^*(k,m)\right|^2}, \quad k=0,1,\ldots,K-1, \quad (7.13)$$

$$\xi_{\text{sr}}\left[\mathbf{h}(:,m)\right] = \frac{\sum_{k=0}^{K-1}\phi_X(k,m)}{\sum_{k=0}^{K-1}\phi_X(k,m)\left|\mathbf{h}^H(k,m)\boldsymbol{\rho}_X^*(k,m)\right|^2}. \quad (7.14)$$

An important observation is that the design of a filter that does not distort the desired signal requires the constraint

$$\mathbf{h}^H(k,m)\boldsymbol{\rho}_X^*(k,m) = 1, \ \forall k,m. \quad (7.15)$$

Thus, the speech reduction factor is equal to 1 if there is no distortion and expected to be greater than 1 when distortion occurs.

Another useful performance measure is the speech distortion index (3), (14) defined as

$$\upsilon_{\text{sd}}\left[\mathbf{h}(k,m)\right] = \frac{E\left\{|X_{\text{ld}}(k,m)-X(k,m)|^2\right\}}{\phi_X(k,m)}$$

$$= \left|\mathbf{h}^H(k,m)\boldsymbol{\rho}_X^*(k,m)-1\right|^2, \quad k=0,1,\ldots,K-1 \quad (7.16)$$

in the narrowband case and as

$$\upsilon_{\text{sd}}\left[\mathbf{h}(:,m)\right] = \frac{\sum_{k=0}^{K-1}E\left\{|X_{\text{ld}}(k,m)-X(k,m)|^2\right\}}{\sum_{k=0}^{K-1}\phi_X(k,m)} \quad (7.17)$$

in the fullband case. The speech distortion index is always greater than or equal to 0 and should be upper bounded by 1 for optimal filters; so the higher is its value, the more the desired signal is distorted.

The error signal between the estimated and desired signals at the frequency-bin k for Model 3 is

$$\mathcal{E}(k,m) = \widehat{X}(k,m) - X(k,m) \quad (7.18)$$
$$= \mathbf{h}^H(k,m)\mathbf{y}(k,m) - X(k,m).$$

We can rewrite (7.18) as

$$\mathcal{E}(k,m) = \mathcal{E}_{\text{d}}(k,m) + \mathcal{E}_{\text{r}}(k,m), \quad (7.19)$$

where

$$\mathcal{E}_{\text{d}}(k,m) = X_{\text{ld}}(k,m) - X(k,m)$$
$$= \left[\mathbf{h}^H(k,m)\boldsymbol{\rho}_X^*(k,m)-1\right]X(k,m) \quad (7.20)$$

is the speech distortion due to the complex filter and

$$
\begin{aligned}
\mathcal{E}_{\mathrm{r}}(k, m) &= X'_{\mathrm{ri}}(k, m) + V_{\mathrm{rn}}(k, m) \\
&= \mathbf{h}^H(k, m)\mathbf{x}'(k, m) + \mathbf{h}^H(k, m)\mathbf{v}(k, m)
\end{aligned}
\tag{7.21}
$$

represents the residual interference-plus-noise.

The narrowband MSE is then

$$
\begin{aligned}
J\left[\mathbf{h}(k, m)\right] &= E\left[|\mathcal{E}(k, m)|^2\right] \\
&= J_{\mathrm{d}}\left[\mathbf{h}(k, m)\right] + J_{\mathrm{r}}\left[\mathbf{h}(k, m)\right],
\end{aligned}
\tag{7.22}
$$

where

$$
\begin{aligned}
J_{\mathrm{d}}\left[\mathbf{h}(k, m)\right] &= E\left[|\mathcal{E}_{\mathrm{d}}(k, m)|^2\right] \\
&= E\left[|X_{1\mathrm{d}}(k, m) - X(k, m)|^2\right] \\
&= \phi_X(k, m)\left|\mathbf{h}^H(k, m)\rho_X^*(k, m) - 1\right|^2
\end{aligned}
\tag{7.23}
$$

and

$$
\begin{aligned}
J_{\mathrm{r}}\left[\mathbf{h}(k, m)\right] &= E\left[|\mathcal{E}_{\mathrm{r}}(k, m)|^2\right] \\
&= E\left[|X'_{\mathrm{ri}}(k, m)|^2\right] + E\left[|V_{\mathrm{rn}}(k, m)|^2\right] \\
&= \phi_{X'_{\mathrm{ri}}}(k, m) + \phi_{V_{\mathrm{rn}}}(k, m).
\end{aligned}
\tag{7.24}
$$

For the particular filter $\mathbf{h}(k, m) = \mathbf{i}_{L,1}$, $\forall k, m$, the narrowband MSE is

$$
J\left[\mathbf{i}_{L,1}(k, m)\right] = \phi_V(k, m),
\tag{7.25}
$$

so there is neither noise reduction nor speech distortion. We can now define the narrowband NMSE as

$$
\begin{aligned}
\tilde{J}\left[\mathbf{h}(k, m)\right] &= \frac{J\left[\mathbf{h}(k, m)\right]}{J\left[\mathbf{i}_{L,1}(k, m)\right]} \\
&= \mathrm{iSNR}(k, m) \cdot \upsilon_{\mathrm{sd}}\left[\mathbf{h}(k, m)\right] + \frac{1}{\xi_{\mathrm{nr}}\left[\mathbf{h}(k, m)\right]},
\end{aligned}
\tag{7.26}
$$

where

$$
\upsilon_{\mathrm{sd}}\left[\mathbf{h}(k, m)\right] = \frac{J_{\mathrm{d}}\left[\mathbf{h}(k, m)\right]}{\phi_X(k, m)},
\tag{7.27}
$$

$$
\xi_{\mathrm{nr}}\left[\mathbf{h}(k, m)\right] = \frac{\phi_V(k, m)}{J_{\mathrm{r}}\left[\mathbf{h}(k, m)\right]}.
\tag{7.28}
$$

It is clear that the objective of noise reduction in the frequency domain with the interframe filtering is to find optimal filters $\mathbf{h}(k, m)$ at each frequency-bin k and time-frame m that would either directly minimize $J\left[\mathbf{h}(k, m)\right]$ or minimize $J_{\mathrm{d}}\left[\mathbf{h}(k, m)\right]$ or $J_{\mathrm{r}}\left[\mathbf{h}(k, m)\right]$ subject to some constraints.

7.2   WIENER FILTER

The Wiener filter is easily derived by taking the gradient of the narrowband MSE, $J\left[\mathbf{h}(k,m)\right]$, with respect to $\mathbf{h}^H(k,m)$ and equating the result to zero:

$$\mathbf{h}_W(k,m) = \Phi_{\mathbf{y}}^{-1}(k,m)\Phi_{\mathbf{yx}}(k,m)\mathbf{i}_{L,1}, \tag{7.29}$$

where

$$\Phi_{\mathbf{y}}(k,m) = E\left[\mathbf{y}(k,m)\mathbf{y}^H(k,m)\right] \tag{7.30}$$

is the covariance matrix of $\mathbf{y}(k,m)$ and

$$\Phi_{\mathbf{yx}}(k,m) = E\left[\mathbf{y}(k,m)\mathbf{x}^H(k,m)\right] \tag{7.31}$$

is the cross-correlation matrix between $\mathbf{y}(k,m)$ and $\mathbf{x}(k,m)$. But

$$\Phi_{\mathbf{yx}}(k,m)\mathbf{i}_{L,1} = \phi_X(k,m)\boldsymbol{\rho}_X^*(k,m), \tag{7.32}$$

so that (7.29) becomes

$$\mathbf{h}_W(k,m) = \phi_X(k,m)\Phi_{\mathbf{y}}^{-1}(k,m)\boldsymbol{\rho}_X^*(k,m). \tag{7.33}$$

The Wiener filter can also be rewritten as

$$\begin{aligned}\mathbf{h}_W(k,m) &= \Phi_{\mathbf{y}}^{-1}(k,m)\Phi_{\mathbf{x}}(k,m)\mathbf{i}_{L,1}\\ &= \left[\mathbf{I}_L - \Phi_{\mathbf{y}}^{-1}(k,m)\Phi_{\mathbf{v}}(k,m)\right]\mathbf{i}_{L,1}.\end{aligned} \tag{7.34}$$

We know that

$$\Phi_{\mathbf{y}}(k,m) = \phi_X(k,m)\boldsymbol{\rho}_X^*(k,m)\boldsymbol{\rho}_X^T(k,m) + \Phi_{\text{in}}(k,m). \tag{7.35}$$

Determining the inverse of $\Phi_{\mathbf{y}}(k,m)$ from (7.35) with the Woodbury's identity, we get

$$\Phi_{\mathbf{y}}^{-1}(k,m) = \Phi_{\text{in}}^{-1}(k,m) - \frac{\Phi_{\text{in}}^{-1}(k,m)\boldsymbol{\rho}_X^*(k,m)\boldsymbol{\rho}_X^T(k,m)\Phi_{\text{in}}^{-1}(k,m)}{\phi_X^{-1}(k,m) + \boldsymbol{\rho}_X^T(k,m)\Phi_{\text{in}}^{-1}(k,m)\boldsymbol{\rho}_X^*(k,m)}. \tag{7.36}$$

Substituting this result into (7.33) leads to another interesting formulation of the Wiener filter:

$$\mathbf{h}_W(k,m) = \frac{\phi_X(k,m)\Phi_{\text{in}}^{-1}(k,m)\boldsymbol{\rho}_X^*(k,m)}{1 + \lambda_{\max}(k,m)}, \tag{7.37}$$

that we can rewrite as

$$\mathbf{h}_W(k,m) = \frac{\Phi_{\text{in}}^{-1}(k,m)\Phi_{\mathbf{y}}(k,m) - \mathbf{I}_L}{1 - L + \text{tr}\left[\Phi_{\text{in}}^{-1}(k,m)\Phi_{\mathbf{y}}(k,m)\right]}\mathbf{i}_{L,1}. \tag{7.38}$$

Using (7.37), we find that the narrowband output SNR is

$$
\begin{aligned}
\text{oSNR}\left[\mathbf{h}_W(k, m)\right] &= \lambda_{\max}(k, m) \\
&= \text{tr}\left[\Phi_{\text{in}}^{-1}(k, m)\Phi_{\mathbf{y}}(k, m)\right] - L
\end{aligned}
\tag{7.39}
$$

and the narrowband speech distortion index is a clear function of this narrowband output SNR:

$$
\upsilon_{\text{sd}}\left[\mathbf{h}_W(k, m)\right] = \frac{1}{\left\{1 + \text{oSNR}\left[\mathbf{h}_W(k, m)\right]\right\}^2}.
\tag{7.40}
$$

Interestingly, the higher is the value of $\text{oSNR}\left[\mathbf{h}_W(k, m)\right]$ (i.e., by increasing the number of inter-frames), the less the desired signal is distorted with the Wiener filter at frequency-bin k.

Clearly,

$$
\text{oSNR}\left[\mathbf{h}_W(k, m)\right] \geq \text{iSNR}(k, m),
\tag{7.41}
$$

since the Wiener filter maximizes the narrowband output SNR. It is of great interest to observe that the two filters $\mathbf{h}_W(k, m)$ and $\mathbf{h}_{\max}(k, m)$ are equivalent up to a scaling factor.

With the Wiener filter, the narrowband noise reduction factor is

$$
\begin{aligned}
\xi_{\text{nr}}\left[\mathbf{h}_W(k, m)\right] &= \frac{[1 + \lambda_{\max}(k, m)]^2}{\text{iSNR}(k, m) \cdot \lambda_{\max}(k, m)} \\
&\geq \left[1 + \frac{1}{\lambda_{\max}(k, m)}\right]^2.
\end{aligned}
\tag{7.42}
$$

Using (7.40) and (7.42) in (7.26), we find the minimum NMSE (MNMSE):

$$
\widetilde{J}\left[\mathbf{h}_W(k, m)\right] = \frac{\text{iSNR}(k, m)}{1 + \lambda_{\max}(k, m)}.
\tag{7.43}
$$

7.3   MVDR FILTER

Remarkably, we can derive a very useful and practical MVDR when the interframe correlation is exploited by minimizing the MSE of the residual interference-plus-noise, $J_{\text{r}}\left[\mathbf{h}(k, m)\right]$, with the constraint that the desired signal is not distorted. Mathematically, this is equivalent to

$$
\min_{\mathbf{h}(k, m)} \mathbf{h}^H(k, m)\Phi_{\text{in}}(k, m)\mathbf{h}(k, m) \quad \text{subject to} \quad \mathbf{h}^H(k, m)\boldsymbol{\rho}_X^*(k, m) = 1,
\tag{7.44}
$$

for which the solution is

$$
\begin{aligned}
\mathbf{h}_{\text{MVDR}}(k, m) &= \frac{\phi_X(k, m)\Phi_{\text{in}}^{-1}(k, m)\boldsymbol{\rho}_X^*(k, m)}{\lambda_{\max}(k, m)} \\
&= \frac{\Phi_{\text{in}}^{-1}(k, m)\Phi_{\mathbf{y}}(k, m) - \mathbf{I}_L}{\text{tr}\left[\Phi_{\text{in}}^{-1}(k, m)\Phi_{\mathbf{y}}(k, m)\right] - L}\mathbf{i}_{L,1}.
\end{aligned}
\tag{7.45}
$$

Alternatively, we can express the MVDR as

$$\mathbf{h}_{\mathrm{MVDR}}(k, m) = \frac{\Phi_{\mathbf{y}}^{-1}(k, m)\boldsymbol{\rho}_X^*(k, m)}{\boldsymbol{\rho}_X^T(k, m)\Phi_{\mathbf{y}}^{-1}(k, m)\boldsymbol{\rho}_X^*(k, m)}. \tag{7.46}$$

The Wiener and MVDR filters are simply related as follows

$$\mathbf{h}_{\mathrm{W}}(k, m) = C(k, m)\mathbf{h}_{\mathrm{MVDR}}(k, m), \tag{7.47}$$

where

$$C(k, m) = \frac{\lambda_{\max}(k, m)}{1 + \lambda_{\max}(k, m)}. \tag{7.48}$$

Here again, the two filters, $\mathbf{h}_{\mathrm{W}}(k, m)$ and $\mathbf{h}_{\mathrm{MVDR}}(k, m)$, are equivalent up to a scaling factor. From a narrowband point of view, this scaling is not significant, but from a fullband point of view, it can be important since speech signals are broadband in nature. Indeed, it can easily be verified that this scaling factor affects the fullband output SNRs and fullband speech distortion indices. While the narrowband output SNRs of the Wiener and MVDR filters are the same, the fullband output SNRs are not because of the scaling factor.

It is clear that we always have

$$\mathrm{oSNR}\left[\mathbf{h}_{\mathrm{MVDR}}(k, m)\right] = \mathrm{oSNR}\left[\mathbf{h}_{\mathrm{W}}(k, m)\right], \tag{7.49}$$

$$\upsilon_{\mathrm{sd}}\left[\mathbf{h}_{\mathrm{MVDR}}(k, m)\right] = 0, \tag{7.50}$$

$$\xi_{\mathrm{sr}}\left[\mathbf{h}_{\mathrm{MVDR}}(k, m)\right] = 1, \tag{7.51}$$

$$\xi_{\mathrm{nr}}\left[\mathbf{h}_{\mathrm{MVDR}}(k, m)\right] = \frac{\lambda_{\max}(k, m)}{\mathrm{iSNR}(k, m)} \leq \xi_{\mathrm{nr}}\left[\mathbf{h}_{\mathrm{W}}(k, m)\right], \tag{7.52}$$

and

$$1 \geq \widetilde{J}\left[\mathbf{h}_{\mathrm{MVDR}}(k, m)\right] = \frac{\mathrm{iSNR}(k, m)}{\lambda_{\max}(k, m)} \geq \widetilde{J}\left[\mathbf{h}_{\mathrm{W}}(k, m)\right]. \tag{7.53}$$

7.4   TRADEOFF FILTER

In the tradeoff approach, we try to compromise between noise reduction and speech distortion. Here, we minimize the speech distortion index with the constraint that the noise reduction factor is equal to a positive value that is greater than 1. Mathematically, this is equivalent to

$$\min_{\mathbf{h}(k,m)} J_{\mathrm{d}}\left[\mathbf{h}(k, m)\right] \quad \text{subject to} \quad J_{\mathrm{r}}\left[\mathbf{h}(k, m)\right] = \beta\phi_V(k, m), \tag{7.54}$$

where $0 < \beta < 1$ to insure that we get some noise reduction. By using a Lagrange multiplier, $\mu \geq 0$, to adjoin the constraint to the cost function, we easily deduce the tradeoff filter:

$$\begin{aligned}
\mathbf{h}_{\mathrm{T},\mu}(k, m) &= \phi_X(k, m)\left[\phi_X(k, m)\boldsymbol{\rho}_X^*(k, m)\boldsymbol{\rho}_X^T(k, m) + \mu\Phi_{\mathrm{in}}(k, m)\right]^{-1}\boldsymbol{\rho}_X^*(k, m) \\
&= \frac{\phi_X(k, m)\Phi_{\mathrm{in}}^{-1}(k, m)\boldsymbol{\rho}_X^*(k, m)}{\mu + \lambda_{\max}(k, m)},
\end{aligned} \tag{7.55}$$

where the Lagrange multiplier, μ, satisfies $J_r\left[\mathbf{h}_{T,\mu}(k,m)\right] = \beta\phi_V(k,m)$. However, in practice, it is not easy to determine the optimal μ. Therefore, when this parameter is chosen in an ad-hoc way, we can see that for

- $\mu = 1, \mathbf{h}_{T,1}(k,m) = \mathbf{h}_W(k,m)$, which is the Wiener filter;

- $\mu = 0, \mathbf{h}_{T,0}(k,m) = \mathbf{h}_{MVDR}(k,m)$, which is the MVDR filter;

- $\mu > 1$, results in a filter with low residual noise at the expense of high speech distortion;

- $\mu < 1$, results in a filter with high residual noise and low speech distortion.

Again, we observe here as well that the tradeoff and Wiener filters are equivalent up to a scaling factor. As a result, the narrowband output SNR with the tradeoff filter is obviously the same as the narrowband output SNR with the Wiener filter, i.e.,

$$\text{oSNR}\left[\mathbf{h}_{T,\mu}(k,m)\right] = \lambda_{\max}(k,m) \tag{7.56}$$

and does not depend on μ. However, the narrowband speech distortion index is now both a function of the variable μ and the narrowband output SNR:

$$\upsilon_{sd}\left[\mathbf{h}_{T,\mu}(k,m)\right] = \frac{\mu^2}{[\mu + \lambda_{\max}(k,m)]^2}. \tag{7.57}$$

From (7.57), we observe how μ can affect the desired signal.

Since the Wiener and MVDR filters are particular cases of the tradeoff filter, it is then useful to study the fullband output SNR and the fullband speech distortion index of the tradeoff filter, which both depend on the variable μ.

Using (7.55) in (7.8), we find that the fullband output SNR is

$$\text{oSNR}\left[\mathbf{h}_{T,\mu}(:,m)\right] = \frac{\sum_{k=0}^{K-1} \dfrac{\phi_X(k,m)\lambda_{\max}^2(k,m)}{[\mu + \lambda_{\max}(k,m)]^2}}{\sum_{k=0}^{K-1} \dfrac{\phi_X(k,m)\lambda_{\max}(k,m)}{[\mu + \lambda_{\max}(k,m)]^2}}. \tag{7.58}$$

We propose the following.

Property 7.1   The fullband output SNR of the tradeoff filter is an increasing function of the parameter μ.

Proof. Indeed, using the proof given in (43) by simply replacing integrals by sums or referring to the proof of Property 6.4 in Section 6.5, we find that

$$\frac{d\,\text{oSNR}\left[\mathbf{h}_{T,\mu}(:,m)\right]}{d\mu} \geq 0, \tag{7.59}$$

proving that the fullband output SNR is increasing when μ is increasing. $\square$

From Property 7.1, we deduce that the MVDR filter gives the smallest fullband output SNR, which is

$$\text{oSNR}\left[\mathbf{h}_{\text{T},0}(:,m)\right] = \frac{\sum_{k=0}^{K-1}\phi_X(k,m)}{\sum_{k=0}^{K-1}\dfrac{\phi_X(k,m)}{\lambda_{\max}(k,m)}}. \tag{7.60}$$

We give another interesting property.

Property 7.2   We have

$$\lim_{\mu\to\infty}\text{oSNR}\left[\mathbf{h}_{\text{T},\mu}(:,m)\right] = \frac{\sum_{k=0}^{K-1}\phi_X(k,m)\lambda_{\max}^2(k,m)}{\sum_{k=0}^{K-1}\phi_X(k,m)\lambda_{\max}(k,m)} \leq \sum_{k=0}^{K-1}\lambda_{\max}(k,m). \tag{7.61}$$

Proof. Easy to show from (7.58). □

While the fullband output SNR is upper bounded, it is easy to show that the fullband noise reduction factor and fullband speech reduction factor are not. So when μ goes to infinity, so are $\xi_{\text{nr}}\left[\mathbf{h}_{\text{T},\mu}(:,m)\right]$ and $\xi_{\text{sr}}\left[\mathbf{h}_{\text{T},\mu}(:,m)\right]$.

The fullband speech distortion index is

$$\upsilon_{\text{sd}}\left[\mathbf{h}_{\text{T},\mu}(:,m)\right] = \frac{\sum_{k=0}^{K-1}\dfrac{\phi_X(k,m)\mu^2}{[\mu+\lambda_{\max}(k,m)]^2}}{\sum_{k=0}^{K-1}\phi_X(k,m)}. \tag{7.62}$$

Property 7.3   The fullband speech distortion index of the tradeoff filter is an increasing function of the parameter μ.

Proof. It is straightforward to verify that

$$\frac{d\upsilon_{\text{sd}}\left[\mathbf{h}_{\text{T},\mu}(:,m)\right]}{d\mu} \geq 0, \tag{7.63}$$

which ends the proof. □

It is clear that

$$0 \leq \upsilon_{\text{sd}}\left[\mathbf{h}_{\text{T},\mu}(:,m)\right] \leq 1, \ \forall\mu \geq 0. \tag{7.64}$$

Therefore, as μ increases, the fullband output SNR increases at the price of more distortion to the desired signal.

Property 7.4   With the tradeoff filter, $\mathbf{h}_{T,\mu}(k, m)$, the fullband output SNR is always greater than or equal to the fullband input SNR, i.e., $\text{oSNR}\left[\mathbf{h}_{T,\mu}(:, m)\right] \geq \text{iSNR}(m), \ \forall \mu \geq 0$.

Proof. We know that

$$\lambda_{\max}(k, m) \geq \text{iSNR}(k, m), \tag{7.65}$$

which implies that

$$\sum_{k=0}^{K-1} \phi_V(k, m) \frac{\text{iSNR}(k, m)}{\lambda_{\max}(k, m)} \leq \sum_{k=0}^{K-1} \phi_V(k, m) \tag{7.66}$$

and, hence,

$$\text{oSNR}\left[\mathbf{h}_{T,0}(:, m)\right] = \frac{\sum_{k=0}^{K-1} \phi_X(k, m)}{\sum_{k=0}^{K-1} \phi_V(k, m) \frac{\text{iSNR}(k, m)}{\lambda_{\max}(k, m)}} \geq \frac{\sum_{k=0}^{K-1} \phi_X(k, m)}{\sum_{k=0}^{K-1} \phi_V(k, m)} = \text{iSNR}(m). \tag{7.67}$$

But from Proposition 7.1, we have

$$\text{oSNR}\left[\mathbf{h}_{T,\mu}(:, m)\right] \geq \text{oSNR}\left[\mathbf{h}_{T,0}(:, m)\right], \ \forall \mu \geq 0, \tag{7.68}$$

as a result,

$$\text{oSNR}\left[\mathbf{h}_{T,\mu}(:, m)\right] \geq \text{iSNR}(m), \ \forall \mu \geq 0, \tag{7.69}$$

which completes the proof.   □

To end this section, let us mention that the decision-directed method (18), which is a reliable estimator of the narrowband input SNR, will fit very well with the proposed algorithms since this estimator implicitly assumes that the successive frames are correlated.

7.5   LINEARLY CONSTRAINED MINIMUM VARIANCE (LCMV) FILTER

We can derive a linearly constrained minimum variance (LCMV) filter (22), (24), which can handle more than one linear constraint, by exploiting the structure of the noise signal (see Section 4.3, Chapter 4):

$$\mathbf{v}(k, m) = V(k, m)\boldsymbol{\rho}_V^*(k, m) + \mathbf{v}'(k, m). \tag{7.70}$$

Our problem this time is the following. We wish to perfectly recover our desired signal, $X(k, m)$, and completely remove the correlated components, $V(k, m)\rho_V^*(k, m)$. Thus, the two constraints can be put together in a matrix form as

$$\mathbf{C}^H(k, m)\mathbf{h}(k, m) = \mathbf{i}_1, \tag{7.71}$$

where

$$\mathbf{C}(k, m) = \left[\begin{array}{cc} \rho_X^*(k, m) & \rho_V^*(k, m) \end{array} \right] \tag{7.72}$$

is our constraint matrix of size $L \times 2$. Then, our optimal filter is obtained by minimizing the energy at the filter output, with the constraints that the correlated noise components are cancelled and the desired speech is preserved, i.e.,

$$\mathbf{h}_{\text{LCMV}}(k, m) = \arg \min_{\mathbf{h}(k,m)} \mathbf{h}^H(k, m)\Phi_\mathbf{y}(k, m)\mathbf{h}(k, m)$$
$$\text{subject to} \quad \mathbf{C}^H(k, m)\mathbf{h}(k, m) = \mathbf{i}_1. \tag{7.73}$$

The solution to (7.73) is given by

$$\mathbf{h}_{\text{LCMV}}(k, m) = \Phi_\mathbf{y}^{-1}(k, m)\mathbf{C}(k, m)\left[\mathbf{C}^H(k, m)\Phi_\mathbf{y}^{-1}(k, m)\mathbf{C}(k, m) \right]^{-1} \mathbf{i}_1. \tag{7.74}$$

By developing (7.74), it can easily be shown that the LCMV can be written as a function of the MVDR:

$$\mathbf{h}_{\text{LCMV}}(k, m) = \frac{1}{1 - |\varpi(k, m)|^2}\mathbf{h}_{\text{MVDR}}(k, m) - \frac{|\varpi(k, m)|^2}{1 - |\varpi(k, m)|^2}\mathbf{t}(k, m), \tag{7.75}$$

where

$$|\varpi(k, m)|^2 = \frac{\left| \rho_X^T(k, m)\Phi_\mathbf{y}^{-1}(k, m)\rho_V^*(k, m) \right|^2}{\left[\rho_X^T(k, m)\Phi_\mathbf{y}^{-1}(k, m)\rho_X^*(k, m) \right]\left[\rho_V^T(k, m)\Phi_\mathbf{y}^{-1}(k, m)\rho_V^*(k, m) \right]}, \tag{7.76}$$

with $0 \leq |\varpi(k, m)|^2 \leq 1$, $\mathbf{h}_{\text{MVDR}}(k, m)$ is defined in (7.46), and

$$\mathbf{t}(k, m) = \frac{\Phi_\mathbf{y}^{-1}(k, m)\rho_V^*(k, m)}{\rho_X^T(k, m)\Phi_\mathbf{y}^{-1}(k, m)\rho_V^*(k, m)}. \tag{7.77}$$

We observe from (7.75) that when $|\varpi(k, m)|^2$ tends to 0, the LCMV filter tends to the MVDR filter; however, when $|\varpi(k, m)|^2$ tends to 1, we have no solution since we have conflicting requirements.

Obviously, we always have

$$\text{oSNR}\left[\mathbf{h}_{\text{LCMV}}(k, m)\right] \leq \text{oSNR}\left[\mathbf{h}_{\text{MVDR}}(k, m)\right], \tag{7.78}$$
$$\upsilon_{\text{sd}}\left[\mathbf{h}_{\text{LCMV}}(k, m)\right] = 0, \tag{7.79}$$
$$\xi_{\text{sr}}\left[\mathbf{h}_{\text{LCMV}}(k, m)\right] = 1, \tag{7.80}$$

and

$$\xi_{nr}\left[\mathbf{h}_{LCMV}(k,m)\right] \leq \xi_{nr}\left[\mathbf{h}_{MVDR}(k,m)\right] \leq \xi_{nr}\left[\mathbf{h}_{W}(k,m)\right]. \tag{7.81}$$

The LCMV filter is able to remove all the correlated noise but at the price that its overall noise reduction is lower than that of the MVDR filter.

CHAPTER 8

Optimal Filters with Model 4

In this chapter, we study noise reduction with Model 4. This is the most general linear model as the interframe correlation is taken into account as well as the widely linear filtering technique (34), (38).

8.1 PERFORMANCE MEASURES

Performance measures given in this section are directly derived from Chapter 3 and Section 4.4, Chapter 4.

To quantify the level of noise remaining at the output of the FIR filter, we define the narrowband output SNR as

$$
\begin{aligned}
\mathrm{oSNR}\left[\widetilde{\underline{\mathbf{h}}}(k, m)\right] &= \frac{\phi_{X_{\mathrm{ld}}}(k, m)}{\phi_{X'_{\mathrm{ri}}}(k, m) + \phi_{V_{\mathrm{rn}}}(k, m)} \\
&= \frac{\phi_X(k, m)\left|\widetilde{\underline{\mathbf{h}}}^H(k, m)\boldsymbol{\varrho}_X^*(k, m)\right|^2}{\widetilde{\underline{\mathbf{h}}}^H(k, m)\Phi_{\mathrm{in}}(k, m)\widetilde{\underline{\mathbf{h}}}(k, m)}, \quad k = 0, 1, \ldots, K-1,
\end{aligned}
\tag{8.1}
$$

where

$$
\Phi_{\mathrm{in}}(k, m) = \Phi_{\widetilde{\underline{\mathbf{x}}}'}(k, m) + \Phi_{\widetilde{\underline{\mathbf{v}}}}(k, m)
\tag{8.2}
$$

is the interference-plus-noise covariance matrix. For the particular filter $\widetilde{\underline{\mathbf{h}}}(k, m) = \mathbf{i}_{2L,1}$ (identity filter), where $\mathbf{i}_{2L,1}$ is the first column of the identity matrix $\mathbf{I}_{2L}$ (of size $2L \times 2L$), we have

$$
\mathrm{oSNR}\left[\mathbf{i}_{2L,1}(k, m)\right] = \mathrm{iSNR}(k, m).
\tag{8.3}
$$

In the previous scenario, the narrowband SNR cannot be improved.

Now, let us define the quantity

$$
\begin{aligned}
\mathrm{oSNR}_{\max}(k, m) &= \mathrm{tr}\left[\Phi_{\mathrm{in}}^{-1}(k, m)\Phi_{\mathbf{x}_{\mathrm{d}}}(k, m)\right] \\
&= \phi_X(k, m)\boldsymbol{\varrho}_X^T(k, m)\Phi_{\mathrm{in}}^{-1}(k, m)\boldsymbol{\varrho}_X^*(k, m).
\end{aligned}
\tag{8.4}
$$

This quantity corresponds to the maximum eigenvalue, $\widetilde{\lambda}_{\max}(k, m)$, of the matrix $\Phi_{\mathrm{in}}^{-1}(k, m)\Phi_{\mathbf{x}_{\mathrm{d}}}(k, m)$. It also corresponds to the maximum output SNR since the filter, $\widetilde{\underline{\mathbf{h}}}_{\max}(k, m)$, that maximizes $\mathrm{oSNR}\left[\widetilde{\underline{\mathbf{h}}}(k, m)\right]$ [eq. (8.1)] is the maximum eigenvector of $\Phi_{\mathrm{in}}^{-1}(k, m)\Phi_{\mathbf{x}_{\mathrm{d}}}(k, m)$ for which its corresponding eigenvalue is $\widetilde{\lambda}_{\max}(k, m)$. As a result, we have

$$
\mathrm{oSNR}\left[\widetilde{\underline{\mathbf{h}}}(k, m)\right] \leq \mathrm{oSNR}_{\max}(k, m) = \widetilde{\lambda}_{\max}(k, m), \quad \forall \widetilde{\underline{\mathbf{h}}}(k, m)
\tag{8.5}
$$

and

$$\text{oSNR}_{\max}(k,m) = \text{oSNR}\left[\widetilde{\underline{\mathbf{h}}}_{\max}(k,m)\right] \geq \text{oSNR}\left[\mathbf{i}_{2L,1}(k,m)\right] = \text{iSNR}(k,m). \tag{8.6}$$

We define the fullband output SNR at time-frame m as

$$\text{oSNR}\left[\widetilde{\underline{\mathbf{h}}}(:,m)\right] = \frac{\sum_{k=0}^{K-1} \phi_X(k,m) \left|\widetilde{\underline{\mathbf{h}}}^H(k,m)\boldsymbol{\varrho}_X^*(k,m)\right|^2}{\sum_{k=0}^{K-1} \widetilde{\underline{\mathbf{h}}}^H(k,m)\boldsymbol{\Phi}_{\text{in}}(k,m)\widetilde{\underline{\mathbf{h}}}(k,m)} \tag{8.7}$$

and it can be verified that (4)

$$\text{oSNR}\left[\widetilde{\underline{\mathbf{h}}}(:,m)\right] \leq \sum_{k=0}^{K-1} \text{oSNR}\left[\widetilde{\underline{\mathbf{h}}}(k,m)\right]. \tag{8.8}$$

As a result,

$$\text{oSNR}\left[\widetilde{\underline{\mathbf{h}}}(:,m)\right] \leq \sum_{k=0}^{K-1} \text{oSNR}_{\max}(k,m). \tag{8.9}$$

The noise reduction factor (3), (14) quantifies the amount of noise that is rejected by the filter. The narrowband and fullband noise reduction factors are then

$$\xi_{\text{nr}}\left[\widetilde{\underline{\mathbf{h}}}(k,m)\right] = \frac{\phi_V(k,m)}{\phi_{X_{\text{ri}}'}(k,m) + \phi_{V_{\text{rn}}}(k,m)}$$

$$= \frac{\phi_V(k,m)}{\widetilde{\underline{\mathbf{h}}}^H(k,m)\boldsymbol{\Phi}_{\text{in}}(k,m)\widetilde{\underline{\mathbf{h}}}(k,m)}, \quad k = 0, 1, \ldots, K-1, \tag{8.10}$$

$$\xi_{\text{nr}}\left[\widetilde{\underline{\mathbf{h}}}(:,m)\right] = \frac{\sum_{k=0}^{K-1} \phi_V(k,m)}{\sum_{k=0}^{K-1} \widetilde{\underline{\mathbf{h}}}^H(k,m)\boldsymbol{\Phi}_{\text{in}}(k,m)\widetilde{\underline{\mathbf{h}}}(k,m)}. \tag{8.11}$$

The noise reduction factors are expected to be lower bounded by 1 for optimal filters. So the more the noise is reduced, the higher are the values of the noise reduction factors.

In the same manner, we define the narrowband and fullband speech reduction factors as

$$\xi_{\text{sr}}\left[\widetilde{\underline{\mathbf{h}}}(k,m)\right] = \frac{\phi_X(k,m)}{\phi_{X_{\text{ld}}}(k,m)}$$

$$= \frac{1}{\left|\widetilde{\underline{\mathbf{h}}}^H(k,m)\boldsymbol{\varrho}_X^*(k,m)\right|^2}, \quad k = 0, 1, \ldots, K-1, \tag{8.12}$$

$$\xi_{\text{sr}}\left[\widetilde{\underline{\mathbf{h}}}(:,m)\right] = \frac{\sum_{k=0}^{K-1} \phi_X(k,m)}{\sum_{k=0}^{K-1} \phi_X(k,m) \left|\widetilde{\underline{\mathbf{h}}}^H(k,m)\boldsymbol{\varrho}_X^*(k,m)\right|^2}. \tag{8.13}$$

An important observation is that the design of a filter that does not distort the desired signal requires the constraint

$$\widetilde{\underline{\mathbf{h}}}^{H}(k, m)\boldsymbol{\varrho}_{X}^{*}(k, m) = 1, \ \forall k, m. \tag{8.14}$$

Thus, the speech reduction factor is equal to 1 if there is no distortion and expected to be greater than 1 when distortion occurs.

Another useful performance measure is the speech distortion index (3), (14) defined as

$$
\begin{aligned}
\upsilon_{\mathrm{sd}}\left[\widetilde{\underline{\mathbf{h}}}(k, m)\right] &= \frac{E\left\{|X_{\mathrm{ld}}(k, m) - X(k, m)|^{2}\right\}}{\phi_{X}(k, m)} \\
&= \left|\widetilde{\underline{\mathbf{h}}}^{H}(k, m)\boldsymbol{\varrho}_{X}^{*}(k, m) - 1\right|^{2}, \ k = 0, 1, \ldots, K - 1
\end{aligned} \tag{8.15}
$$

in the narrowband case and as

$$\upsilon_{\mathrm{sd}}\left[\widetilde{\underline{\mathbf{h}}}(:, m)\right] = \frac{\sum_{k=0}^{K-1} E\left\{|X_{\mathrm{ld}}(k, m) - X(k, m)|^{2}\right\}}{\sum_{k=0}^{K-1} \phi_{X}(k, m)} \tag{8.16}$$

in the fullband case. The speech distortion index is always greater than or equal to 0 and should be upper bounded by 1 for optimal filters; so the higher is its value, the more the desired signal is distorted.

The error signal between the estimated and desired signals at the frequency-bin k for Model 4 is

$$
\begin{aligned}
\mathcal{E}(k, m) &= \widehat{X}(k, m) - X(k, m) \\
&= \widetilde{\underline{\mathbf{h}}}^{H}(k, m)\widetilde{\underline{\mathbf{y}}}(k, m) - X(k, m).
\end{aligned} \tag{8.17}
$$

We can rewrite (8.17) as

$$\mathcal{E}(k, m) = \mathcal{E}_{\mathrm{d}}(k, m) + \mathcal{E}_{\mathrm{r}}(k, m), \tag{8.18}$$

where

$$
\begin{aligned}
\mathcal{E}_{\mathrm{d}}(k, m) &= X_{\mathrm{ld}}(k, m) - X(k, m) \\
&= \left[\widetilde{\underline{\mathbf{h}}}^{H}(k, m)\boldsymbol{\varrho}_{X}^{*}(k, m) - 1\right] X(k, m)
\end{aligned} \tag{8.19}
$$

is the speech distortion due to the complex filter and

$$
\begin{aligned}
\mathcal{E}_{\mathrm{r}}(k, m) &= X'_{\mathrm{ri}}(k, m) + V_{\mathrm{rn}}(k, m) \\
&= \widetilde{\underline{\mathbf{h}}}^{H}(k, m)\widetilde{\underline{\mathbf{x}}}'(k, m) + \widetilde{\underline{\mathbf{h}}}^{H}(k, m)\widetilde{\underline{\mathbf{v}}}(k, m)
\end{aligned} \tag{8.20}
$$

represents the residual interference-plus-noise.

The narrowband MSE is then

$$
\begin{aligned}
J\left[\underline{\widetilde{\mathbf{h}}}(k,m)\right] &= E\left[|\mathcal{E}(k,m)|^2\right] \\
&= J_{\mathrm{d}}\left[\underline{\widetilde{\mathbf{h}}}(k,m)\right] + J_{\mathrm{r}}\left[\underline{\widetilde{\mathbf{h}}}(k,m)\right],
\end{aligned}
\tag{8.21}
$$

where

$$
\begin{aligned}
J_{\mathrm{d}}\left[\underline{\widetilde{\mathbf{h}}}(k,m)\right] &= E\left[|\mathcal{E}_{\mathrm{d}}(k,m)|^2\right] \\
&= E\left[|X_{\mathrm{ld}}(k,m) - X(k,m)|^2\right] \\
&= \phi_X(k,m)\left|\underline{\widetilde{\mathbf{h}}}^H(k,m)\varrho_X^*(k,m) - 1\right|^2
\end{aligned}
\tag{8.22}
$$

and

$$
\begin{aligned}
J_{\mathrm{r}}\left[\underline{\widetilde{\mathbf{h}}}(k,m)\right] &= E\left[|\mathcal{E}_{\mathrm{r}}(k,m)|^2\right] \\
&= E\left[|X_{\mathrm{ri}}'(k,m)|^2\right] + E\left[|V_{\mathrm{rn}}(k,m)|^2\right] \\
&= \phi_{X_{\mathrm{ri}}'}(k,m) + \phi_{V_{\mathrm{rn}}}(k,m).
\end{aligned}
\tag{8.23}
$$

For the particular filter $\underline{\widetilde{\mathbf{h}}}(k,m) = \mathbf{i}_{2L,1}$, $\forall k, m$, the narrowband MSE is

$$
J\left[\mathbf{i}_{2L,1}(k,m)\right] = \phi_V(k,m),
\tag{8.24}
$$

so there is neither noise reduction nor speech distortion. We can now define the narrowband NMSE as

$$
\begin{aligned}
\widetilde{J}\left[\underline{\widetilde{\mathbf{h}}}(k,m)\right] &= \frac{J\left[\underline{\widetilde{\mathbf{h}}}(k,m)\right]}{J\left[\mathbf{i}_{2L,1}(k,m)\right]} \\
&= \mathrm{iSNR}(k,m) \cdot \upsilon_{\mathrm{sd}}\left[\underline{\widetilde{\mathbf{h}}}(k,m)\right] + \frac{1}{\xi_{\mathrm{nr}}\left[\underline{\widetilde{\mathbf{h}}}(k,m)\right]},
\end{aligned}
\tag{8.25}
$$

where

$$
\upsilon_{\mathrm{sd}}\left[\underline{\widetilde{\mathbf{h}}}(k,m)\right] = \frac{J_{\mathrm{d}}\left[\underline{\widetilde{\mathbf{h}}}(k,m)\right]}{\phi_X(k,m)},
\tag{8.26}
$$

$$
\xi_{\mathrm{nr}}\left[\underline{\widetilde{\mathbf{h}}}(k,m)\right] = \frac{\phi_V(k,m)}{J_{\mathrm{r}}\left[\underline{\widetilde{\mathbf{h}}}(k,m)\right]}.
\tag{8.27}
$$

The clear objective of noise reduction in the frequency domain with the interframe and widely linear filtering is to find optimal filters $\underline{\widetilde{\mathbf{h}}}(k,m)$ at each frequency-bin k and time-frame m that would either directly minimize $J\left[\underline{\widetilde{\mathbf{h}}}(k,m)\right]$ or minimize $J_{\mathrm{d}}\left[\underline{\widetilde{\mathbf{h}}}(k,m)\right]$ or $J_{\mathrm{r}}\left[\underline{\widetilde{\mathbf{h}}}(k,m)\right]$, subject to some constraint.

8.2   WIENER FILTER

The Wiener filter is obtained by taking the gradient of the narrowband MSE, $J\left[\underline{\widetilde{\mathbf{h}}}(k,m)\right]$, with respect to $\underline{\widetilde{\mathbf{h}}}^{H}(k,m)$ and equating the result to zero:

$$\underline{\widetilde{\mathbf{h}}}_{\mathrm{W}}(k,m) = \Phi_{\underline{\widetilde{\mathbf{y}}}}^{-1}(k,m)\Phi_{\underline{\widetilde{\mathbf{y}}}\underline{\widetilde{\mathbf{x}}}}(k,m)\mathbf{i}_{2L,1}, \tag{8.28}$$

where

$$\Phi_{\underline{\widetilde{\mathbf{y}}}}(k,m) = E\left[\underline{\widetilde{\mathbf{y}}}(k,m)\underline{\widetilde{\mathbf{y}}}^{H}(k,m)\right] \tag{8.29}$$

is the covariance matrix of $\underline{\widetilde{\mathbf{y}}}(k,m)$ and

$$\begin{aligned}
\Phi_{\underline{\widetilde{\mathbf{y}}}\underline{\widetilde{\mathbf{x}}}}(k,m) &= E\left[\underline{\widetilde{\mathbf{y}}}(k,m)\underline{\widetilde{\mathbf{x}}}^{H}(k,m)\right] \\
&= \Phi_{\underline{\widetilde{\mathbf{x}}}}(k,m) \tag{8.30}
\end{aligned}$$

is the cross-correlation matrix between $\underline{\widetilde{\mathbf{y}}}(k,m)$ and $\underline{\widetilde{\mathbf{x}}}(k,m)$. But

$$\Phi_{\underline{\widetilde{\mathbf{y}}}\underline{\widetilde{\mathbf{x}}}}(k,m)\mathbf{i}_{2L,1} = \phi_{X}(k,m)\boldsymbol{\varrho}_{X}^{*}(k,m), \tag{8.31}$$

so that (8.28) becomes

$$\underline{\widetilde{\mathbf{h}}}_{\mathrm{W}}(k,m) = \phi_{X}(k,m)\Phi_{\underline{\widetilde{\mathbf{y}}}}^{-1}(k,m)\boldsymbol{\varrho}_{X}^{*}(k,m). \tag{8.32}$$

The Wiener filter can also be rewritten as

$$\begin{aligned}
\underline{\widetilde{\mathbf{h}}}_{\mathrm{W}}(k,m) &= \Phi_{\underline{\widetilde{\mathbf{y}}}}^{-1}(k,m)\Phi_{\underline{\widetilde{\mathbf{x}}}}(k,m)\mathbf{i}_{2L,1} \\
&= \left[\mathbf{I}_{2L} - \Phi_{\underline{\widetilde{\mathbf{y}}}}^{-1}(k,m)\Phi_{\underline{\widetilde{\mathbf{v}}}}(k,m)\right]\mathbf{i}_{2L,1}. \tag{8.33}
\end{aligned}$$

We know that

$$\Phi_{\underline{\widetilde{\mathbf{y}}}}(k,m) = \phi_{X}(k,m)\boldsymbol{\varrho}_{X}^{*}(k,m)\boldsymbol{\varrho}_{X}^{T}(k,m) + \Phi_{\mathrm{in}}(k,m). \tag{8.34}$$

Determining the inverse of $\Phi_{\underline{\widetilde{\mathbf{y}}}}(k,m)$ from (8.34) with the Woodbury's identity, we get

$$\Phi_{\underline{\widetilde{\mathbf{y}}}}^{-1}(k,m) = \Phi_{\mathrm{in}}^{-1}(k,m) - \frac{\Phi_{\mathrm{in}}^{-1}(k,m)\boldsymbol{\varrho}_{X}^{*}(k,m)\boldsymbol{\varrho}_{X}^{T}(k,m)\Phi_{\mathrm{in}}^{-1}(k,m)}{\phi_{X}^{-1}(k,m) + \boldsymbol{\varrho}_{X}^{T}(k,m)\Phi_{\mathrm{in}}^{-1}(k,m)\boldsymbol{\varrho}_{X}^{*}(k,m)}. \tag{8.35}$$

Substituting this result into (8.32) leads to another interesting formulation of the Wiener filter:

$$\underline{\widetilde{\mathbf{h}}}_{\mathrm{W}}(k,m) = \frac{\phi_{X}(k,m)\Phi_{\mathrm{in}}^{-1}(k,m)\boldsymbol{\varrho}_{X}^{*}(k,m)}{1 + \underline{\lambda}_{\mathrm{max}}(k,m)}, \tag{8.36}$$

that we can rewrite as

$$\underline{\widetilde{\mathbf{h}}}_{\mathrm{W}}(k, m) = \frac{\Phi_{\mathrm{in}}^{-1}(k, m)\Phi_{\underline{\widetilde{\mathbf{y}}}}(k, m) - \mathbf{I}_{2L}}{1 - 2L + \mathrm{tr}\left[\Phi_{\mathrm{in}}^{-1}(k, m)\Phi_{\underline{\widetilde{\mathbf{y}}}}(k, m)\right]}\mathbf{i}_{2L,1}. \tag{8.37}$$

Using (8.36), we find that the narrowband output SNR is

$$\begin{aligned}
\mathrm{oSNR}\left[\underline{\widetilde{\mathbf{h}}}_{\mathrm{W}}(k, m)\right] &= \widetilde{\lambda}_{\max}(k, m) \\
&= \mathrm{tr}\left[\Phi_{\mathrm{in}}^{-1}(k, m)\Phi_{\underline{\widetilde{\mathbf{y}}}}(k, m)\right] - 2L
\end{aligned} \tag{8.38}$$

and the narrowband speech distortion index is a clear function of this narrowband output SNR:

$$\upsilon_{\mathrm{sd}}\left[\underline{\widetilde{\mathbf{h}}}_{\mathrm{W}}(k, m)\right] = \frac{1}{\left\{1 + \mathrm{oSNR}\left[\underline{\widetilde{\mathbf{h}}}_{\mathrm{W}}(k, m)\right]\right\}^2}. \tag{8.39}$$

Interestingly, the higher is the value of $\mathrm{oSNR}\left[\underline{\widetilde{\mathbf{h}}}_{\mathrm{W}}(k, m)\right]$, the less the desired signal is distorted with the Wiener filter at frequency-bin k.

Clearly,

$$\mathrm{oSNR}\left[\underline{\widetilde{\mathbf{h}}}_{\mathrm{W}}(k, m)\right] \geq \mathrm{iSNR}(k, m), \tag{8.40}$$

since the Wiener filter maximizes the narrowband output SNR. It is of great interest to observe that the two filters, $\underline{\widetilde{\mathbf{h}}}_{\mathrm{W}}(k, m)$ and $\underline{\widetilde{\mathbf{h}}}_{\max}(k, m)$, are equivalent up to a scaling factor.

With the Wiener filter, the narrowband noise reduction factor is

$$\begin{aligned}
\xi_{\mathrm{nr}}\left[\underline{\widetilde{\mathbf{h}}}_{\mathrm{W}}(k, m)\right] &= \frac{\left[1 + \widetilde{\lambda}_{\max}(k, m)\right]^2}{\mathrm{iSNR}(k, m) \cdot \widetilde{\lambda}_{\max}(k, m)} \\
&\geq \left[1 + \frac{1}{\widetilde{\lambda}_{\max}(k, m)}\right]^2.
\end{aligned} \tag{8.41}$$

Using (8.39) and (8.41) in (8.25), we find the minimum NMSE (MNMSE):

$$\widetilde{J}\left[\underline{\widetilde{\mathbf{h}}}_{\mathrm{W}}(k, m)\right] = \frac{\mathrm{iSNR}(k, m)}{1 + \widetilde{\lambda}_{\max}(k, m)}. \tag{8.42}$$

8.3   MVDR FILTER

With Model 4, we can also derive an MVDR filter by minimizing the MSE of the residual interference-plus-noise, $J_{\mathrm{r}}\left[\underline{\widetilde{\mathbf{h}}}(k, m)\right]$, with the constraint that the desired signal is not distorted. Mathematically, this is equivalent to

$$\min_{\underline{\widetilde{\mathbf{h}}}(k,m)} \underline{\widetilde{\mathbf{h}}}^H(k, m)\Phi_{\mathrm{in}}(k, m)\underline{\widetilde{\mathbf{h}}}(k, m) \quad \text{subject to} \quad \underline{\widetilde{\mathbf{h}}}^H(k, m)\varrho_X^*(k, m) = 1, \tag{8.43}$$

for which the solution is

$$
\begin{aligned}
\widetilde{\underline{\mathbf{h}}}_{\mathrm{MVDR}}(k, m) &= \frac{\phi_X(k, m)\Phi_{\mathrm{in}}^{-1}(k, m)\varrho_X^*(k, m)}{\widetilde{\lambda}_{\max}(k, m)} \\
&= \frac{\Phi_{\mathrm{in}}^{-1}(k, m)\Phi_{\widetilde{\mathbf{y}}}(k, m) - \mathbf{I}_{2L}}{\mathrm{tr}\left[\Phi_{\mathrm{in}}^{-1}(k, m)\Phi_{\widetilde{\mathbf{y}}}(k, m)\right] - 2L}\mathbf{i}_{2L,1}.
\end{aligned}
\tag{8.44}
$$

Obviously, we can rewrite the MVDR as

$$
\widetilde{\underline{\mathbf{h}}}_{\mathrm{MVDR}}(k, m) = \frac{\Phi_{\widetilde{\mathbf{y}}}^{-1}(k, m)\varrho_X^*(k, m)}{\varrho_X^T(k, m)\Phi_{\widetilde{\mathbf{y}}}^{-1}(k, m)\varrho_X^*(k, m)}.
\tag{8.45}
$$

The Wiener and MVDR filters are simply related as follows

$$
\widetilde{\underline{\mathbf{h}}}_{\mathrm{W}}(k, m) = \widetilde{\underline{C}}(k, m)\widetilde{\underline{\mathbf{h}}}_{\mathrm{MVDR}}(k, m),
\tag{8.46}
$$

where

$$
\widetilde{\underline{C}}(k, m) = \frac{\widetilde{\lambda}_{\max}(k, m)}{1 + \widetilde{\lambda}_{\max}(k, m)}.
\tag{8.47}
$$

Here again the two filters, $\widetilde{\underline{\mathbf{h}}}_{\mathrm{W}}(k, m)$ and $\widetilde{\underline{\mathbf{h}}}_{\mathrm{MVDR}}(k, m)$, are equivalent up to a scaling factor. From a narrowband point of view, this scaling is not significant, but from a fullband point of view, it can be important since speech signals are broadband in nature. Indeed, it can easily be verified that this scaling factor affects the fullband output SNRs and fullband speech distortion indices. While the narrowband output SNRs of the Wiener and MVDR filters are the same, the fullband output SNRs are not because of the scaling factor.

It is clear that we always have

$$
\begin{aligned}
\mathrm{oSNR}\left[\widetilde{\underline{\mathbf{h}}}_{\mathrm{MVDR}}(k, m)\right] &= \mathrm{oSNR}\left[\widetilde{\underline{\mathbf{h}}}_{\mathrm{W}}(k, m)\right], &\tag{8.48} \\
\upsilon_{\mathrm{sd}}\left[\widetilde{\underline{\mathbf{h}}}_{\mathrm{MVDR}}(k, m)\right] &= 0, &\tag{8.49} \\
\xi_{\mathrm{sr}}\left[\widetilde{\underline{\mathbf{h}}}_{\mathrm{MVDR}}(k, m)\right] &= 1, &\tag{8.50} \\
\xi_{\mathrm{nr}}\left[\widetilde{\underline{\mathbf{h}}}_{\mathrm{MVDR}}(k, m)\right] &= \frac{\widetilde{\lambda}_{\max}(k, m)}{\mathrm{iSNR}(k, m)} \leq \xi_{\mathrm{nr}}\left[\widetilde{\underline{\mathbf{h}}}_{\mathrm{W}}(k, m)\right], &\tag{8.51}
\end{aligned}
$$

and

$$
1 \geq \widetilde{J}\left[\widetilde{\underline{\mathbf{h}}}_{\mathrm{MVDR}}(k, m)\right] = \frac{\mathrm{iSNR}(k, m)}{\widetilde{\lambda}_{\max}(k, m)} \geq \widetilde{J}\left[\widetilde{\underline{\mathbf{h}}}_{\mathrm{W}}(k, m)\right].
\tag{8.52}
$$

8.4   TRADEOFF FILTER

The tradeoff approach tries to compromise between noise reduction and speech distortion. Here, the speech distortion index is minimized with the constraint that the noise reduction factor is equal to a positive value that is greater than 1. Mathematically, this is equivalent to

$$\min_{\widetilde{\mathbf{h}}(k,m)} J_d\left[\widetilde{\mathbf{h}}(k,m)\right] \quad \text{subject to} \quad J_r\left[\widetilde{\mathbf{h}}(k,m)\right] = \beta\phi_V(k,m), \tag{8.53}$$

where $0 < \beta < 1$ to insure that we get some noise reduction. By using a Lagrange multiplier, $\mu \geq 0$, to adjoin the constraint to the cost function, we deduce the tradeoff filter:

$$\begin{aligned}
\widetilde{\mathbf{h}}_{T,\mu}(k,m) &= \phi_X(k,m)\left[\phi_X(k,m)\boldsymbol{\varrho}_X^*(k,m)\boldsymbol{\varrho}_X^T(k,m) + \mu\Phi_{\text{in}}(k,m)\right]^{-1}\boldsymbol{\varrho}_X^*(k,m) \\
&= \frac{\phi_X(k,m)\Phi_{\text{in}}^{-1}(k,m)\boldsymbol{\varrho}_X^*(k,m)}{\mu + \widetilde{\lambda}_{\max}(k,m)},
\end{aligned} \tag{8.54}$$

where the Lagrange multiplier, μ, satisfies $J_r\left[\widetilde{\mathbf{h}}_{T,\mu}(k,m)\right] = \beta\phi_V(k,m)$. However, in practice, it is not easy to determine the optimal μ. Therefore, when this parameter is chosen in an ad-hoc way, we can see that for

- $\mu = 1, \widetilde{\mathbf{h}}_{T,1}(k,m) = \widetilde{\mathbf{h}}_W(k,m)$, which is the Wiener filter;

- $\mu = 0, \widetilde{\mathbf{h}}_{T,0}(k,m) = \widetilde{\mathbf{h}}_{MVDR}(k,m)$, which is the MVDR filter;

- $\mu > 1$, results in a filter with low residual noise at the expense of high speech distortion;

- $\mu < 1$, results in a filter with high residual noise and low speech distortion.

Again, we observe here as well that the tradeoff and Wiener filters are equivalent up to a scaling factor. As a result, the narrowband output SNR with the tradeoff filter is obviously the same as the narrowband output SNR with the Wiener filter, i.e.,

$$\text{oSNR}\left[\widetilde{\mathbf{h}}_{T,\mu}(k,m)\right] = \widetilde{\lambda}_{\max}(k,m) \tag{8.55}$$

and does not depend on μ. However, the narrowband speech distortion index is now both a function of the variable μ and the narrowband output SNR:

$$\upsilon_{sd}\left[\widetilde{\mathbf{h}}_{T,\mu}(k,m)\right] = \frac{\mu^2}{\left[\mu + \widetilde{\lambda}_{\max}(k,m)\right]^2}. \tag{8.56}$$

From (8.56), we observe how μ can affect the desired signal.

Since the Wiener and MVDR filters are particular cases of the tradeoff filter, it is then useful to study the fullband output SNR and the fullband speech distortion index of the tradeoff filter, which both depend on the variable μ.

Using (8.54) in (8.7), we find that the fullband output SNR is

$$\text{oSNR}\left[\widetilde{\underline{\mathbf{h}}}_{\text{T},\mu}(:,m)\right] = \frac{\sum_{k=0}^{K-1} \dfrac{\phi_X(k,m)\widetilde{\underline{\lambda}}_{\max}^2(k,m)}{\left[\mu + \widetilde{\underline{\lambda}}_{\max}(k,m)\right]^2}}{\sum_{k=0}^{K-1} \dfrac{\phi_X(k,m)\widetilde{\underline{\lambda}}_{\max}(k,m)}{\left[\mu + \widetilde{\underline{\lambda}}_{\max}(k,m)\right]^2}}. \tag{8.57}$$

We propose the following.

Property 8.1 The fullband output SNR of the tradeoff filter is an increasing function of the parameter μ.

Proof. Indeed, using the proof given in (43) by simply replacing integrals by sums or referring to the proof of Property 6.4 in Section 6.5, we find that

$$\frac{d\text{oSNR}\left[\widetilde{\underline{\mathbf{h}}}_{\text{T},\mu}(:,m)\right]}{d\mu} \geq 0, \tag{8.58}$$

proving that the fullband output SNR is increasing when μ is increasing. □

From Property 8.1, we deduce that the MVDR filter gives the smallest fullband output SNR, which is

$$\text{oSNR}\left[\widetilde{\underline{\mathbf{h}}}_{\text{T},0}(:,m)\right] = \frac{\sum_{k=0}^{K-1} \phi_X(k,m)}{\sum_{k=0}^{K-1} \dfrac{\phi_X(k,m)}{\widetilde{\underline{\lambda}}_{\max}(k,m)}}. \tag{8.59}$$

We give another interesting property.

Property 8.2 We have

$$\lim_{\mu \to \infty} \text{oSNR}\left[\widetilde{\underline{\mathbf{h}}}_{\text{T},\mu}(:,m)\right] = \frac{\sum_{k=0}^{K-1} \phi_X(k,m)\widetilde{\underline{\lambda}}_{\max}^2(k,m)}{\sum_{k=0}^{K-1} \phi_X(k,m)\widetilde{\underline{\lambda}}_{\max}(k,m)} \leq \sum_{k=0}^{K-1} \widetilde{\underline{\lambda}}_{\max}(k,m). \tag{8.60}$$

Proof. Easy to show from (8.57). □

While the fullband output SNR is upper bounded, it is easy to show that the fullband noise reduction factor and fullband speech reduction factor are not. So when μ goes to infinity, so are $\xi_{nr}\left[\widetilde{\underline{\mathbf{h}}}_{T,\mu}(:, m)\right]$ and $\xi_{sr}\left[\widetilde{\underline{\mathbf{h}}}_{T,\mu}(:, m)\right]$.

The fullband speech distortion index is

$$
\upsilon_{sd}\left[\widetilde{\underline{\mathbf{h}}}_{T,\mu}(:, m)\right] = \frac{\sum_{k=0}^{K-1} \dfrac{\phi_X(k, m)\mu^2}{\left[\mu + \widetilde{\underline{\lambda}}_{max}(k, m)\right]^2}}{\sum_{k=0}^{K-1} \phi_X(k, m)}.
\tag{8.61}
$$

Property 8.3   The fullband speech distortion index of the tradeoff filter is an increasing function of the parameter μ.

Proof. It is straightforward to verify that

$$
\frac{d\upsilon_{sd}\left[\widetilde{\underline{\mathbf{h}}}_{T,\mu}(:, m)\right]}{d\mu} \geq 0,
\tag{8.62}
$$

which ends the proof.   □

It is clear that

$$
0 \leq \upsilon_{sd}\left[\widetilde{\underline{\mathbf{h}}}_{T,\mu}(:, m)\right] \leq 1, \ \forall \mu \geq 0.
\tag{8.63}
$$

Therefore, as μ increases, the fullband output SNR increases at the price of more distortion to the desired signal.

Property 8.4   With the tradeoff filter, $\widetilde{\underline{\mathbf{h}}}_{T,\mu}(k, m)$, the fullband output SNR is always greater than or equal to the fullband input SNR, i.e., $\text{oSNR}\left[\widetilde{\underline{\mathbf{h}}}_{T,\mu}(:, m)\right] \geq \text{iSNR}(m), \ \forall \mu \geq 0$.

Proof. We know that

$$
\widetilde{\underline{\lambda}}_{max}(k, m) \geq \text{iSNR}(k, m),
\tag{8.64}
$$

which implies that

$$
\sum_{k=0}^{K-1} \phi_V(k, m) \frac{\text{iSNR}(k, m)}{\widetilde{\underline{\lambda}}_{max}(k, m)} \leq \sum_{k=0}^{K-1} \phi_V(k, m)
\tag{8.65}
$$

and, hence,

$$\text{oSNR}\left[\widetilde{\underline{\mathbf{h}}}_{T,0}(:,m)\right] = \frac{\sum_{k=0}^{K-1} \phi_X(k,m)}{\sum_{k=0}^{K-1} \phi_V(k,m)\dfrac{\text{iSNR}(k,m)}{\widetilde{\lambda}_{\max}(k,m)}} \geq \frac{\sum_{k=0}^{K-1} \phi_X(k,m)}{\sum_{k=0}^{K-1} \phi_V(k,m)} = \text{iSNR}(m). \quad (8.66)$$

But from Proposition 8.1, we have

$$\text{oSNR}\left[\widetilde{\underline{\mathbf{h}}}_{T,\mu}(:,m)\right] \geq \text{oSNR}\left[\widetilde{\underline{\mathbf{h}}}_{T,0}(:,m)\right], \ \forall \mu \geq 0, \quad (8.67)$$

as a result,

$$\text{oSNR}\left[\widetilde{\underline{\mathbf{h}}}_{T,\mu}(:,m)\right] \geq \text{iSNR}(m), \ \forall \mu \geq 0, \quad (8.68)$$

which completes the proof.   □

8.5   LCMV FILTER

We can derive an LCMV filter (22), (24), which can handle more than one linear constraint, by exploiting the structure of the noise signal (see Section 4.4, Chapter 4):

$$\widetilde{\underline{\mathbf{v}}}(k,m) = V(k,m)\boldsymbol{\varrho}_V^*(k,m) + \underline{\widetilde{\mathbf{v}}}(k,m). \quad (8.69)$$

Our problem this time is the following. We wish to perfectly recover our desired signal, $X(k,m)$, and completely remove the correlated components, $V(k,m)\boldsymbol{\varrho}_V^*(k,m)$. Thus, the two constraints can be put together in a matrix form as

$$\widetilde{\underline{\mathbf{C}}}^H(k,m)\widetilde{\underline{\mathbf{h}}}(k,m) = \mathbf{i}_1, \quad (8.70)$$

where

$$\widetilde{\underline{\mathbf{C}}}(k,m) = \begin{bmatrix} \boldsymbol{\varrho}_X^*(k,m) & \boldsymbol{\varrho}_V^*(k,m) \end{bmatrix} \quad (8.71)$$

is our constraint matrix of size $2L \times 2$. Then, our optimal filter is obtained by minimizing the energy at the filter output, with the constraints that the correlated noise components are cancelled and the desired speech is preserved, i.e.,

$$\widetilde{\underline{\mathbf{h}}}_{\text{LCMV}}(k,m) = \arg\min_{\widetilde{\underline{\mathbf{h}}}(k,m)} \widetilde{\underline{\mathbf{h}}}^H(k,m)\Phi_{\widetilde{\underline{\mathbf{y}}}}(k,m)\widetilde{\underline{\mathbf{h}}}(k,m)$$

$$\text{subject to} \quad \widetilde{\underline{\mathbf{C}}}^H(k,m)\widetilde{\underline{\mathbf{h}}}(k,m) = \mathbf{i}_1. \quad (8.72)$$

The solution to (8.72) is given by

$$\widetilde{\underline{\mathbf{h}}}_{\text{LCMV}}(k,m) = \Phi_{\widetilde{\underline{\mathbf{y}}}}^{-1}(k,m)\widetilde{\underline{\mathbf{C}}}(k,m)\left[\widetilde{\underline{\mathbf{C}}}^H(k,m)\Phi_{\widetilde{\underline{\mathbf{y}}}}^{-1}(k,m)\widetilde{\underline{\mathbf{C}}}(k,m)\right]^{-1} \mathbf{i}_1. \quad (8.73)$$

By developing (8.73), it can easily be shown that the LCMV can be written as a function of the MVDR:

$$\widetilde{\underline{\mathbf{h}}}_{\text{LCMV}}(k, m) = \frac{1}{1 - |\widetilde{\underline{\varpi}}(k, m)|^2} \widetilde{\underline{\mathbf{h}}}_{\text{MVDR}}(k, m) - \frac{|\widetilde{\underline{\varpi}}(k, m)|^2}{1 - |\widetilde{\underline{\varpi}}(k, m)|^2} \widetilde{\underline{\mathbf{t}}}(k, m), \tag{8.74}$$

where

$$|\widetilde{\underline{\varpi}}(k, m)|^2 = \frac{\left| \boldsymbol{\varrho}_X^T(k, m) \Phi_{\widetilde{\underline{\mathbf{y}}}}^{-1}(k, m) \boldsymbol{\varrho}_V^*(k, m) \right|^2}{\left[\boldsymbol{\varrho}_X^T(k, m) \Phi_{\widetilde{\underline{\mathbf{y}}}}^{-1}(k, m) \boldsymbol{\varrho}_X^*(k, m) \right] \left[\boldsymbol{\varrho}_V^T(k, m) \Phi_{\widetilde{\underline{\mathbf{y}}}}^{-1}(k, m) \boldsymbol{\varrho}_V^*(k, m) \right]}, \tag{8.75}$$

with $0 \le |\widetilde{\underline{\varpi}}(k, m)|^2 \le 1$, $\widetilde{\underline{\mathbf{h}}}_{\text{MVDR}}(k, m)$ is defined in (8.45), and

$$\widetilde{\underline{\mathbf{t}}}(k, m) = \frac{\Phi_{\widetilde{\underline{\mathbf{y}}}}^{-1}(k, m) \boldsymbol{\varrho}_V^*(k, m)}{\boldsymbol{\varrho}_X^T(k, m) \Phi_{\widetilde{\underline{\mathbf{y}}}}^{-1}(k, m) \boldsymbol{\varrho}_V^*(k, m)}. \tag{8.76}$$

We observe from (8.74) that when $|\widetilde{\underline{\varpi}}(k, m)|^2$ tends to 0, the LCMV filter tends to the MVDR filter; however, when $|\widetilde{\underline{\varpi}}(k, m)|^2$ tends to 1, we have no solution since we have conflicting requirements.

Obviously, we always have

$$\text{oSNR}\left[\widetilde{\underline{\mathbf{h}}}_{\text{LCMV}}(k, m) \right] \le \text{oSNR}\left[\widetilde{\underline{\mathbf{h}}}_{\text{MVDR}}(k, m) \right], \tag{8.77}$$

$$\upsilon_{\text{sd}}\left[\widetilde{\underline{\mathbf{h}}}_{\text{LCMV}}(k, m) \right] = 0, \tag{8.78}$$

$$\xi_{\text{sr}}\left[\widetilde{\underline{\mathbf{h}}}_{\text{LCMV}}(k, m) \right] = 1, \tag{8.79}$$

and

$$\xi_{\text{nr}}\left[\widetilde{\underline{\mathbf{h}}}_{\text{LCMV}}(k, m) \right] \le \xi_{\text{nr}}\left[\widetilde{\underline{\mathbf{h}}}_{\text{MVDR}}(k, m) \right] \le \xi_{\text{nr}}\left[\widetilde{\underline{\mathbf{h}}}_{\text{W}}(k, m) \right]. \tag{8.80}$$

The LCMV filter is able to remove all the correlated noise but at the price that its overall noise reduction is lower than that of the MVDR filter.

CHAPTER 9

Experimental Study

In this chapter, we study the performance of the developed frequency-domain single-channel noise reduction algorithms with simulations. Among a great number of the discussed algorithms, only the most typical and the most promising filters are investigated. The rest will be left to the reader for further exploration. Since we already learned from analytical analyzes presented in Chapter 6 that the gain in SNR from exploiting the circularity characteristics of speech and the additive noise would be marginal (upper-bounded by 3 dB if noise is stationary), the focus here will be placed on the evaluation of the new noise reduction filters with Model 3 in comparison to those conventional approaches with Model 1.

9.1 SETUP AND METRICS

In our experiments, the microphone signal is artificially synthesized by adding computer-generated white Gaussian random noise or pre-recorded real-world noise to a clean speech signal. The clean speech signals were recorded from 22 female and 24 male talkers with a Sennheiser ME-80 Condenser Super Cardioid 50-15000 Hz microphone in the Bell Labs Murray Hill's anechoic chamber in July/August 1989. Each talker provided 2 to 8 minutes of "conversational speech" that is a "story" about anything that came to his/her mind. All recordings were originally digitized at a sampling rate of 48 kHz with 16 bits per sample. They were then downsampled to 8 kHz for our use. In the experiments presented here, we consider only 2 female and 2 male speakers. Each story was cut to have the same length of 60 s. For real-world noise, we only consider car and babble noise. The car noise is fairly stationary but colored with an energy roll-off (approximately 12 dB per octave) towards high frequencies. The babble noise was recorded in the New York Stock Exchange (NYSE). It is not only colored but also non-stationary with mixtures of nearly inaudible voices and sporadic cell phone rings. The noise level is adjusted according to that of the clean speech and a specified input SNR. In the following, if not explicitly stated otherwise, the noise is white Gaussian random noise and the speech source is the first female talker.

In our experiments, the fullband output SNR, oSNR(m), and the fullband speech distortion index, $\upsilon_{\text{sd}}(m)$, are used. But instead of showing these measures on a frame-by-frame basis, we define and present their long-term counterparts. They include the long-term input SNR, the long-term output SNR, and the long-term speech distortion index, given respectively by,

$$\text{iSNR} = \frac{\sum_{m=0}^{M-1} \sum_{k=0}^{K=1} \phi_X(k,m)}{\sum_{m=0}^{M-1} \sum_{k=0}^{K-1} \phi_V(k,m)}, \tag{9.1}$$

$$\tag{9.2}$$

$$\text{oSNR} = \frac{\sum_{m=0}^{M-1} \sum_{k=0}^{K-1} |\rho_1(k,m)|^2 \phi_X(k,m)}{\sum_{m=0}^{M-1} \sum_{k=0}^{K-1} \left[|\rho_2(k,m)|^2 \phi_V(k,m) + \phi_W(k,m) \right]}, \tag{9.3}$$

$$\upsilon_{\text{sd}} = \frac{\sum_{m=0}^{M-1} \sum_{k=0}^{K-1} E\left\{ |X_{\text{ld}}(k,m) - X(k,m)|^2 \right\}}{\sum_{m=0}^{M-1} \sum_{k=0}^{K-1} \phi_X(k,m)}, \tag{9.4}$$

where M is the total number of signal frames. Similar to the relationships between the narrowband and fullband SNRs as in (3.3) and (3.9), we can easily deduce the following relationships for the short-term and long-term SNRs:

$$\text{iSNR} \leq \sum_{m=0}^{M-1} \text{iSNR}(m), \tag{9.5}$$

$$\text{oSNR} \leq \sum_{m=0}^{M-1} \text{oSNR}(m). \tag{9.6}$$

We intend to investigate both the conventional definitions (i.e., those with Model 1, see Section 5.1) and the new definitions (e.g., those with Model 3, see Section 7.1) of the performance measures. Moreover, we will use the PESQ (Perceptual Evaluation of Speech Quality (50)) to obtain an objective assessment of the overall quality of the speech that are enhanced by the investigated noise reduction algorithms.

9.2   ALGORITHM IMPLEMENTATION

The algorithms discussed and developed in this book are all frequency-domain approaches. The STFT is implemented with the Kaiser window and the FFT (fast Fourier transform). The window size K in samples is set to be a power of 2, and there is a 75% overlap between neighboring windows. So every time the window slides over $K/4$ samples. The overlap-add method then is used for signal reconstruction in the time domain to avoid errors caused by circular convolution. This analysis and synthesis procedure is nearly perfect in Matlab, resulting in little distortion in the reconstructed signal if no manipulation is carried out to its frequency-domain representations. The level of distortion varies with K, as shown in Table 9.1.

The window size determines the FFT resolution and also affects the calculated interframe correlations used in Model 3. These two effects can have probably opposite impacts on the performance of the developed algorithms for Model 3, which will be explored in the experiments.

Table 9.1: Accuracy of the frequency analysis and resynthesis procedure used in the presented experiments. The distortion level is measured with reference to the input signal.

FFT Window Size K (samples)	Distortion (dB)
16	-75.9
32	-107.2
64	-90.3
128	-87.5

In all experiments, we use the first 100 frames to compute the initial estimates of $\Phi_\mathbf{y}(k, m)$ and $\Phi_\mathbf{v}(k, m)$ by averaging with a batch method. The rest of the signal frames are then used for performance evaluation. In this process, the estimates of $\Phi_\mathbf{y}(k, m)$ and $\Phi_\mathbf{v}(k, m)$ are recursively updated according to

$$
\begin{aligned}
\Phi_\mathbf{y}(k, m) &= \lambda_y \Phi_\mathbf{y}(k, m - 1) + (1 - \lambda_y)\mathbf{y}(k, m)\mathbf{y}^H(k, m), & (9.7) \\
\Phi_\mathbf{v}(k, m) &= \lambda_v \Phi_\mathbf{v}(k, m - 1) + (1 - \lambda_v)\mathbf{v}(k, m)\mathbf{v}^H(k, m), & (9.8)
\end{aligned}
$$

where $0 < \lambda_y < 1$ and $0 < \lambda_v < 1$ are the forgetting factors. In order to remove the uncertainty of voice activity detection in an otherwise more practical but less rigorous performance comparison, we choose to update the estimates of noise statistics continuously from the noise signal in these simulations.

After $\Phi_\mathbf{y}(k, m)$ and $\Phi_\mathbf{v}(k, m)$ become available at time-frame m, $\Phi_\mathbf{x}(k, m)$ is computed as

$$
\Phi_\mathbf{x}(k, m) = \Phi_\mathbf{y}(k, m) - \Phi_\mathbf{v}(k, m). \tag{9.9}
$$

The normalized interframe correlation vector of speech $\boldsymbol{\rho}_X(k, m)$ is then taken as the transpose of the first row vector of $\Phi_\mathbf{x}(k, m)$ normalized by its first element.

To compute the inverse of $\Phi_\mathbf{y}(k, m)$, as in (7.29) or in (7.46), the technique of regularization is used, so that $\Phi_\mathbf{y}^{-1}(k, m)$ is replaced by

$$
\left\{ \Phi_\mathbf{y}(k, m) + \frac{\delta \cdot \mathrm{tr}\left[\Phi_\mathbf{y}(k, m)\right]}{L}\mathbf{I}_L \right\}^{-1}, \tag{9.10}
$$

where $\delta > 0$ is the regularization factor. It is empirically set as $\delta = 0.01$ in this chapter. When the inverse of $\Phi_{\mathrm{in}}(k, m)$ instead of $\Phi_\mathbf{y}(k, m)$ needs to be calculated, as in (7.38) or in (7.45), the technique of regularization will be similarly applied.

9.3   COMPARISON OF THE WIENER FILTERS WITH MODEL 1 AND MODEL 3

We first show the performance of the traditional Wiener filter with Model 1, which provides a benchmark for studying other noise reduction filters in the rest of this chapter. In this experiment, the long-term input SNR is 0 dB and the window size $K = 64$ samples (corresponding to 8 ms). We set $\lambda_v = 0.96, 0.86$, and 0.72, respectively, and examine how the long-term, fullband performance varies with λ_y. Figure 9.1 plots the results. Using a large λ_y (close to 1), we cannot capture the short-term variation of inherently nonstationary speech signals. But on the other hand with a small λ_y, the sample estimate of the signal variance $\phi_Y(k, m)$ has a large variation due to a limited number of data to do averaging. So the best performance is achieved with a moderate λ_y. An interesting observation is that the output SNR reaches its peak when $\lambda_y = \lambda_v$. The performance of the Wiener filter with $\lambda_y = \lambda_v$ is then also included in Fig. 9.1 for easy comparison. Since $\phi_V(k, m)$ is directly computed and continuously updated from the noise signal in our simulations, the match of λ_y to λ_v leads to the same tracking characteristics for $\phi_Y(k, m)$ and $\phi_V(k, m)$, and hence better noise reduction performances.

The second experiment considers the Wiener filter with Model 3. Again, we fix iSNR = 0 dB, $K = 64$, and $\lambda_y = \lambda_v$, but let L go from 1 up to 8. The results are presented in Fig. 9.2. When L becomes larger, the size of the covariance matrix $\Phi_\mathbf{y}(k, m)$ that needs to be inverted in the Wiener filter with Model 3 grows. As a result, the optimal forgetting factors λ_y and λ_v that produce the best noise reduction performance increase. It is evident that using multiple STFT frames (i.e., Model 3) is helpful. Comparing the best-case scenario of $L = 4$ against that of $L = 1$ (equivalent to Model 1), we see that the output SNR is improved by approximately 2 dB while the level of speech distortion remains almost the same. But when L is larger than 4, the gain in the SNR will no longer significantly increase. At the same time, the performance becomes more sensitive to λ_y and the algorithm becomes more computationally intensive due to a larger size of $\Phi_\mathbf{y}(k, m)$.

The third experiment was designed to reveal how the window size K affects the performance of the Wiener filter with Model 3. We consider iSNR = 0 dB and $L = 4$, and make $\lambda_y = \lambda_v$, but let K (a power of 2) vary from 16 to 128. Figure 9.3 shows the results of this study. When K is small (e.g., 16), the FFT resolution is poor. In this case, increasing K is helpful to improve the performance. But when K moves from 64 to 128, the obtained performance gain becomes very marginal. This is due to the fact that a large K corresponds to a long gap in time between two consecutive frames and hence leads to a weaker interframe correlation.

The fourth and last experiment with the Wiener filters was to examine the benefit of using Model 3 (i.e., multiple frames) under different input SNR conditions. This time, we fix $K = 64$ and again make $\lambda_y = \lambda_v$. We let iSNR be $-10, 0$, or 10 dB, and L be 1 or 4. The results are visualized in Fig. 9.4. It is promising that the gain of using multiple frames is observable over a practically wide range of input SNR's. An interesting discovery is that the gain is greater for a low iSNR than for a high iSNR.

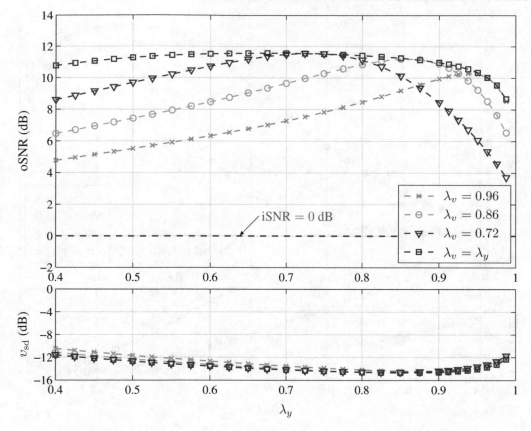

Figure 9.1: Effect of the forgetting factors λ_y and λ_v on the long-term, fullband performance of the traditional Wiener filter with Model 1. The window size is $K = 64$ (8 ms) and iSNR = 0 dB.

Before we conclude this section, there is one thing that needs to be clarified and discussed. It is about which set of performance measures we used for the experiments presented above. On the contrary to ones intuitions, we didn't use the performance measures based on Model 3 (discussed in Section 7.1) for the Wiener filters based on the same model. But instead, we used the conventional definitions on the basis of Model 1 (see Section 5.1) for all the Wiener filters. These two sets of definitions for performance measures differ in the way of treating the interference. The interference is produced by multi-frame processing from the speech components of the previous frames that are uncorrelated with the desired speech in the current frame. So in the traditional Wiener filter with Model 1, the interference is always zero and the two sets of performance measures definitions are equivalent. But in Model 3 that uses multiple frames ($L > 1$), it is more insightful to consider the interference as a part of noise in the new definition of output SNR, or similarly in that of speech distortion measure. While this makes sense, we found that the performance measures with

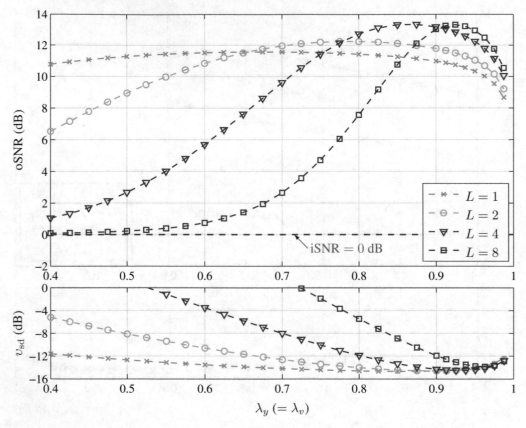

Figure 9.2: Comparison in the long-term, fullband performance of the Wiener filters with Model 3 using various numbers of consecutive STFT frames L. The forgetting factors are $\lambda_y = \lambda_v$, the window size is $K = 64$ (8 ms), and iSNR = 0 dB.

Model 3 are not reliable in practice with the Wiener filters. By definition, the interference vector $\mathbf{x}'(k, m)$ is uncorrelated with $X(k, m)$ and, therefore, the three components of the narrowband error in (7.18) through (7.21) are mutually uncorrelated too. As a result, the narrowband MSE $J\left[\mathbf{h}(k, m)\right]$ is the sum of $J_{\mathrm{d}}\left[\mathbf{h}(k, m)\right]$, $\phi_{X'_{\mathrm{ri}}}(k, m)$, and $\phi_{V_{\mathrm{rn}}}(k, m)$ [see (7.22), (7.23), and (7.24)], which are all non-negative. When the narrowband MSE is minimized by the Wiener filter, the power of the residual interference $\phi_{X'_{\mathrm{ri}}}(k, m)$ will be small (at least smaller than the MSE), and the conventional (Model 1) and new (Model 3) definitions of output SNR (respectively speech distortion) should not be significantly different. But in practice, the estimate of the normalized interframe correlation vector $\boldsymbol{\rho}_X(k, m)$ can never be accurate, and the estimation error leads to a leakage of speech into the interference. Consequently, there remain some correlations between the calculated speech distortion $\mathcal{E}_{\mathrm{d}}(k, m)$ [see (7.20)] and the calculated residual interference $X'_{\mathrm{ri}}(k, m)$

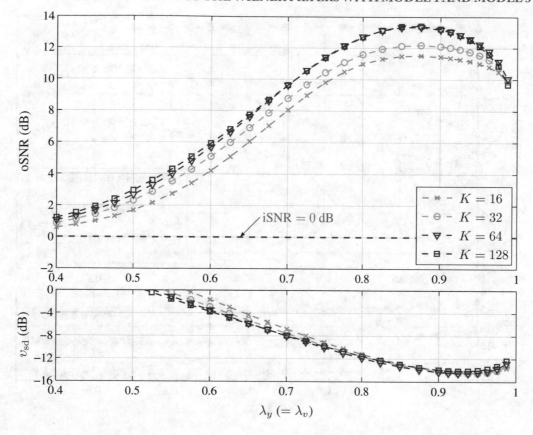

Figure 9.3: Effect of the window size on the long-term, fullband performance of the Wiener filter with Model 3, which uses multiple consecutive STFT frames. The number of frames used in the filters is $L = 4$, the forgetting factors are $\lambda_y = \lambda_v$, and iSNR = 0 dB.

in (7.21). So the decomposition of (7.22) through (7.24) does not hold, and the power of the estimated residual interference can be even greater than the minimum MSE, which makes the definitions on the basis of Model 3 unreliable. This issue is illustrated by the waveforms presented in Fig. 9.5. They are the results of the Wiener filter with Model 3 for which we set $L = 4$, iSNR = 0 dB, $K = 64$, and $\lambda_y = \lambda_v = 0.9$. We see that $x_{\mathrm{ri}}(t)$ is clearly correlated with $x_{\mathrm{fd}}(t)$ and has a large variance. If the definitions with Model 3 are used, oSNR = 0.23 dB and $\upsilon_{\mathrm{sd}} = 11.90$ dB, which do not reflect our perception of the quality of the enhanced signal. The signal of $x_{\mathrm{ri}}(t) + x_{\mathrm{fd}}(t)$ of this Wiener filter looks and sounds similar to the clean speech, and the noise level has been clearly reduced. Using the definition with Model 1, we get that oSNR = 13.19 dB and $\upsilon_{\mathrm{sd}} = -14.30$ dB.

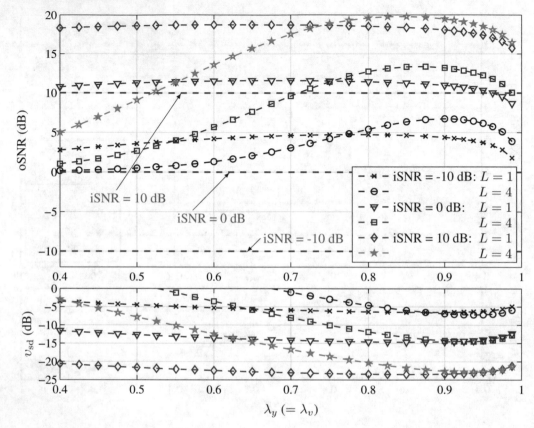

Figure 9.4: Comparison of the traditional Wiener filter with Model 1 ($L = 1$) and the new Wiener filter with Model 3 ($L = 4$) under different input SNR conditions. The forgetting factors are $\lambda_y = \lambda_v$ and the window size is $K = 64$ (8 ms).

9.4   MVDR FILTER WITH MODEL 3

In this section, we study the MVDR filter with Model 3. Note that the performance measures on the basis of Model 3 are used. While the signal decomposition can still be problematic as explained above, the MVDR filter minimizes the variance of the residual interference-plus-noise under the constraint of no distortion in the filtered desired speech signal such that the speech leakage in the interference will be suppressed too.

The first experiment considers some MVDR filters that use different numbers of frames for iSNR = 0 dB and with the window size being set as $K = 64$. Again, we make $\lambda_y = \lambda_v$. Figure 9.6 shows the results. The levels of the speech distortion index of these MVDR filters clearly indicate that the speech distortionless constraint has been met regardless of various values of L. As L increases,

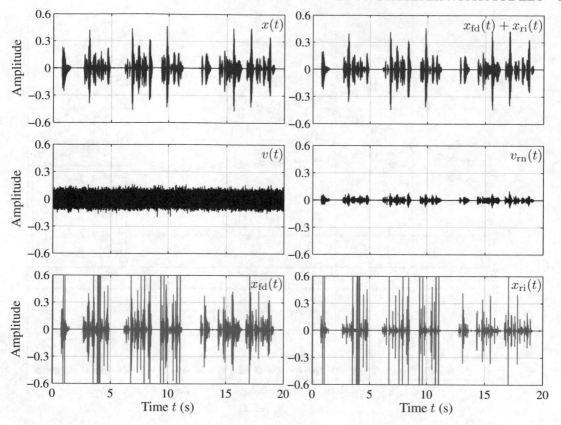

Figure 9.5: Signal waveforms of a noise reduction Wiener filter with Model 3 ($L = 4$) (only the first 20 seconds are shown): the clean speech $x(t)$, the additive Gaussian noise $v(t)$, the filtered desired speech $x_{\mathrm{fd}}(t)$, the residual interference $x_{\mathrm{ri}}(t)$, and the residual noise $v_{\mathrm{rn}}(t)$. The input SNR is iSNR = 0 dB, $K = 64$, and $\lambda_y = \lambda_v = 0.9$.

the optimal λ_y and λ_v become larger, and the best oSNR improves. When L reaches 12, the benefit of further increasing the number vanishes and is strongly outweighed by the cost of increased complexity.

Figure 9.7 presents the results of the experiment that investigates the impact of the window size on the MVDR filter with Model 3. The input SNR is 0 dB and $L = 12$. It is clear that the optimal forgetting factors do not change much with the window size, and $K = 64$ produces the best output SNR (only slightly better than $K = 32$).

Figure 9.8 shows the long-term, fullband performance of the MVDR filter under different input SNR conditions. We consider $L = 12$ and $K = 64$, and make $\lambda_y = \lambda_v$. The speech distortionless constraint has always been satisfied, and the gain in SNR declines as the input SNR increases.

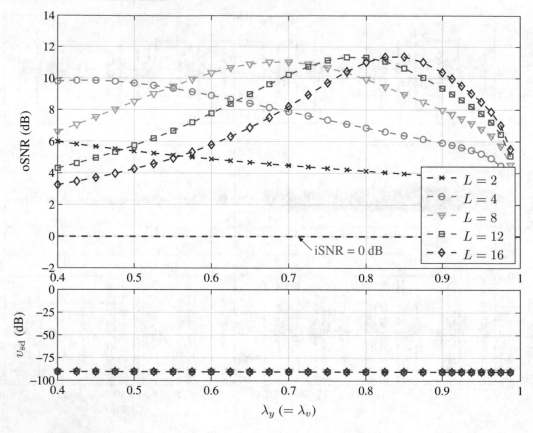

Figure 9.6: Comparison in the long-term, fullband performance of the MVDR filters with Model 3 using various numbers of consecutive STFT frames L. The forgetting factors are $\lambda_y = \lambda_v$, the window size is $K = 64$ (8 ms), and iSNR = 0 dB.

9.5   PERCEPTUAL QUALITY

The conducted research has indicated that the output SNR and the speech distortion index provide a complete and insightful picture of the noise reduction performance. They are closely aligned with our perception of the quality of the enhanced signals in informal listening tests, if the proper set of definitions are applied since each set has some caveats as explained above at the end of Section 9.3. Using the same set of definitions, it has become clear that exploiting interframe correlations (i.e., Model 3) is helpful with either the Wiener or the MVDR filters. But it can give rise to arguments if we compare the performance of the Wiener and MVDR filters (both based on Model 3) using different sets of performance measure definitions. So for this task, we chose to use the PESQ measure, which has been found to have higher correlations than other widely known objective measures, with

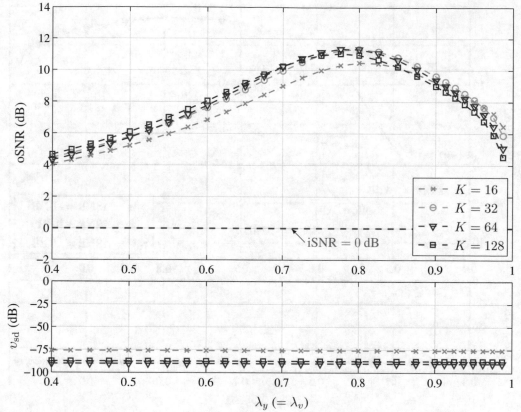

Figure 9.7: Effect of the window size on the long-term, fullband performance of the MVDR filter with Model 3. The number of used frames is $L = 4$, the forgetting factors are $\lambda_y = \lambda_v$, and iSNR = 0 dB.

the subjective ratings of overall quality of enhanced speech signals (27), (40). All 4 talkers were used to find the average PESQ score for each tested condition. Such a raw PESQ MOS (Mean Opinion Score) is then mapped to the PESQ MOS-LQO (Listening Quality Objective) to make a linear connection to subjective MOS using the following mapping function (51):

$$\text{PESQ}_{\text{MOS-LQO}} = 0.999 + \frac{4}{1 + e^{-1.4945 \times \text{PESQ}_{\text{MOS}} + 4.6607}}. \tag{9.11}$$

PESQ MOS-LQO ranges between 1.02 and 4.55.

Figure 9.9 shows the results for the three different noise types. The window size is set as $K = 64$. For the traditional single-frame (Model 1, $L = 1$) and the new multi-frame (Model 3, $L = 4$) Wiener filters, we make that $\lambda_y = \lambda_v = 0.9$. For the MVDR filter with Model 3, we take $L = 12$ and $\lambda_y = \lambda_v = 0.8$. The multi-frame Wiener filter (Model 3) performs always better than

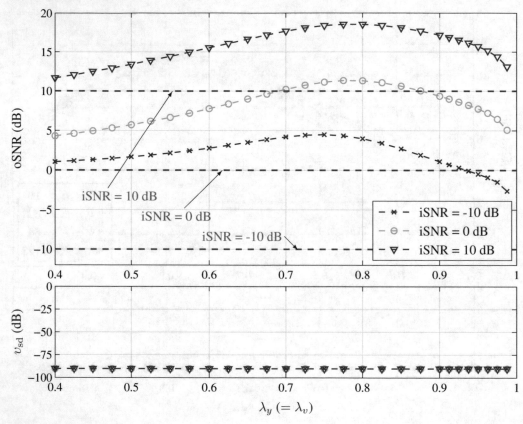

Figure 9.8: Long-term, fullband performance of the MVDR filter with Model 3 under different input SNR conditions. The forgetting factors are $\lambda_y = \lambda_v$, the number of used frames is $L = 12$, and the window size is $K = 64$ (8 ms).

the single-frame counterpart (Model 1) for all noise types. It is noted that the MVDR filter based on Model 3 produces low speech distortion but high residual noise. When the input SNR is low (lower than 10 dB), the high level of the residual noise outweighs speech distortion in the PESQ measure such that the MVDR filter yields lower PESQ scores than the two Wiener filters. On the contrary, when the input SNR gets practically high, speech distortion becomes much easier to be perceived with lower residual noise in the background. Consequently, the MVDR filter has higher PESQ scores than the Wiener filters in those conditions.

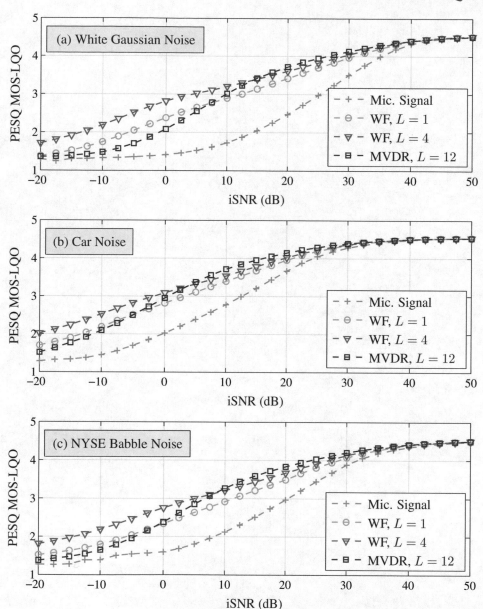

Figure 9.9: Comparison of the traditional single-frame Wiener filter based on Model 1 (WF, $L = 1$), the multi-frame Wiener filter based on Model 3 (WF, $L = 4$), and the Model-3-based MVDR filter (MVDR, $L = 12$) using the PESQ MOS-LQO measure in: (a) white Gaussian noise, (b) car noise, and (c) NYSE babble noise. The window size is $K = 64$. For the two Wiener filters, $\lambda_y = \lambda_v = 0.9$. For the MVDR filter, $\lambda_y = \lambda_v = 0.8$.

Bibliography

[1] P. O. Amblard, M. Gaeta, and J. L. Lacoume, "Statistics for complex variables and signals – Part I: variables," *Elsevier Signal Process.*, vol. 53, pp. 1–13, 1996. DOI: 10.1016/0165-1684(96)00071-0 18

[2] P. O. Amblard, M. Gaeta, and J. L. Lacoume, "Statistics for complex variables and signals – Part II: signals," *Elsevier Signal Process.*, vol. 53, pp. 15–25, 1996. DOI: 10.1016/0165-1684(96)00072-2 18

[3] J. Benesty, J. Chen, Y. Huang, and S. Doclo, "Study of the Wiener filter for noise reduction," in *Speech Enhancement*, J. Benesty, S. Makino, and J. Chen, Eds., Berlin, Germany: Springer-Verlag, 2005, Chapter 2, pp. 9–41. 12, 14, 54, 55, 66, 67

[4] J. Benesty, J. Chen, Y. Huang, and I. Cohen, *Noise Reduction in Speech Processing*. Berlin, Germany: Springer-Verlag, 2009. 5, 11, 12, 13, 21, 30, 31, 46, 54, 66

[5] J. Benesty, J. Chen, and Y. Huang, "On noise reduction in the Karhunen-Loève expansion domain," in *Proc. IEEE ICASSP*, 2009, pp. 25–28. DOI: 10.1109/ICASSP.2009.4959511 21

[6] J. Benesty, J. Chen, and Y. Huang, "On widely linear Wiener and tradeoff filters for noise reduction," *Speech Communication*, vol. 52, pp. 427–439, May 2010. DOI: 10.1016/j.specom.2010.02.003 6, 39, 47

[7] J. Benesty, J. Chen, and Y. Huang, "A widely linear distortionless filter for single-channel noise reduction," *IEEE Signal Process. Lett.*, vol. 17, pp. 469–472, May 2010. DOI: 10.1109/LSP.2010.2043152 6, 44

[8] J. Benesty and J. Chen, *Optimal Time-Domain Noise Reduction Filters – A Theoretical Study*. Berlin, Germany: Springer, 2011. 2

[9] J. Benesty, J. Chen, and Y. Huang, *Speech Enhancement in the Karhunen-Loève Expansion Domain*. San Rafael, CA: Morgan & Claypool Publishers, 2011. 2

[10] M. Berouti, M. Schwartz, and J. Makhoul, "Enhancement of speech corrupted by acoustic noise," in *Proc. IEEE ICASSP*, 1979, pp. 208–211. DOI: 10.1109/ICASSP.1979.1170788 2

[11] S. F. Boll, "Suppression of acoustic noise in speech using spectral subtraction," *IEEE Trans. Acoust., Speech, Signal Process.*, vol. ASSP-27, pp. 113–120, Apr. 1979. DOI: 10.1109/TASSP.1979.1163209 1

[12] V. Bray and M. Valente, "Can omni-directional hearing aids improve speech understanding in noise?" *Audiology Online*, Sept. 24, 2001. Accessed on Jan. 31, 2011, available at `http://www.audiologyonline.com/articles/article_detail.asp?article_id=300`. 1

[13] J. Capon, "High resolution frequency-wavenumber spectrum analysis," *Proc. IEEE*, vol. 57, pp. 1408–1418, Aug. 1969. DOI: 10.1109/PROC.1969.7278 44

[14] J. Chen, J. Benesty, Y. Huang, and S. Doclo, "New insights into the noise reduction Wiener filter," *IEEE Trans. Audio, Speech, Language Process.*, vol. 14, pp. 1218–1234, July 2006. DOI: 10.1109/TSA.2005.860851 12, 14, 54, 55, 66, 67

[15] J. Chen, J. Benesty, and Y. Huang, "Study of the noise-reduction problem in the Karhunen-Loève expansion domain," *IEEE Trans. Audio, Speech, Language Process.*, vol. 17, pp. 787–802, May 2009. DOI: 10.1109/TASL.2009.2014793 21

[16] P. Chevalier, J.-P. Delmas, and A. Oukaci, "Optimal widely linear MVDR beam-forming for noncircular signals," in *Proc. IEEE ICASSP*, 2009, pp. 3573–3576. DOI: 10.1109/ICASSP.2009.4960398 19

[17] I. Cohen, "Relaxed statistical model for speech enhancement and a priori SNR estimation," *IEEE Trans. Speech, Audio Process.*, vol. 13, pp. 870–881, Sept. 2005. DOI: 10.1109/TSA.2005.851940 21

[18] Y. Ephraim and D. Malah, "Speech enhancement using a minimum mean-square error short-time spectral amplitude estimator," *IEEE Trans. Acoust., Speech, Signal Process.*, vol. ASSP-32, pp. 1109–1121, Dec. 1984. DOI: 10.1109/TASSP.1984.1164453 2, 62

[19] Y. Ephraim and D. Malah, "Speech enhancement using a minimum mean-square error log-spectral amplitude estimator," *IEEE Trans. Acoust., Speech, Signal Process.*, vol. ASSP-33, pp. 443–445, Apr. 1985. DOI: 10.1109/TASSP.1985.1164550 2

[20] Y. Ephraim, D. Malah, and B.-H. Juang, "On the application of hidden Markov models for enhancing noisy speech," *IEEE Trans. Acoust., Speech, Signal Process.*, vol. ASSP-37, pp. 1846–1856, Dec. 1989. DOI: 10.1109/29.45532 2

[21] Y. Ephraim and H. L. Van Trees, "A signal subspace approach for speech enhancement," *IEEE Trans. Speech, Audio Process.*, vol. 3, pp. 251–266, July 1995. DOI: 10.1109/89.397090 32

[22] M. Er and A. Cantoni, "Derivative constraints for broad-band element space antenna array processors," *IEEE Trans. Acoust., Speech, Signal Process.*, vol. ASSP-31, pp. 1378–1393, Dec. 1983. DOI: 10.1109/TASSP.1983.1164219 62, 75

[23] J. Eriksson and V. Koivunen, "Complex random vectors and ICA models: identifiability, uniqueness, and separability," *IEEE Trans. Inf. Theory*, vol. 52, pp. 1017–1029, Mar. 2006. DOI: 10.1109/TIT.2005.864440 5, 6

[24] O. Frost, "An algorithm for linearly constrained adaptive array process.," *Proc. IEEE*, vol. 60, pp. 926–935, Jan. 1972. DOI: 10.1109/PROC.1972.8817 62, 75

[25] Y. Hu and P. C. Loizou, "A generalized subspace approach for enhancing speech corrupted by colored noise," *IEEE Trans. Speech, Audio Process.*, vol. 11, pp. 334–341, July 2003. DOI: 10.1109/TSA.2003.814458 32

[26] Y. Hu and P. C. Loizou, "A perceptually motivated approach for speech enhancement," *IEEE Trans. Speech, Audio Process.*, vol. 11, pp. 457–465, Sept. 2003. DOI: 10.1109/TSA.2003.815936 32

[27] Y. Hu and P. C. Loizou, "Evaluation of objective quality measures for speech enhancement," *IEEE Trans. Audio, Speech, Language Process.*, vol. 16, no. 1, pp. 229–238, Jan. 2008. DOI: 10.1109/TASL.2007.911054 87

[28] F. Jabloun and B. Champagne, "Incorporating the human hearing properties in the signal subspace approach for speech enhancement," *IEEE Trans. Speech, Audio Process.*, vol. 11, pp. 700–708, Nov. 2003. DOI: 10.1109/TSA.2003.818031 32

[29] R. T. Lacoss, "Data adaptive spectral analysis methods," *Geophysics*, vol. 36, pp. 661–675, Aug. 1971. DOI: 10.1190/1.1440203 44

[30] J. S. Lim and A. V. Oppenheim, "Enhancement and bandwidth compression of noisy speech," *Proc. IEEE*, vol. 67, pp. 1586–1604, Dec. 1979. DOI: 10.1109/PROC.1979.11540 1, 2

[31] P. C. Loizou, "Speech enhancement based on perceptually motivated Bayesian estimators of the magnitude spectrum," *IEEE Trans. Speech, Audio Process.*, vol. 13, pp. 857–869, Sept. 2005. DOI: 10.1109/TSA.2005.851929 2

[32] P. Loizou, *Speech Enhancement: Theory and Practice*. Boca Raton, FL: CRC Press, 2007. 5

[33] T. Lotter and P. Vary, "Noise reduction by maximum a posteriori spectral amplitude estimation with supergaussian speech modelling," in *Proc. IWAENC*, 2003, pp. 83–86. 2

[34] D. P. Mandic and S. L. Goh, *Complex Valued Nonlinear Adaptive Filters: Noncircularity, Widely Linear and Neural Models*. Wiley, 2009. 5, 6, 18, 24, 35, 65

[35] R. J. McAulay and M. L. Malpass, "Speech enhancement using a soft-decision noise suppression filter," *IEEE Trans. Acoust., Speech, Signal Process.*, vol. ASSP-28, pp. 137–145, Apr. 1980. DOI: 10.1109/TASSP.1980.1163394 1, 2

[36] E. Ollila, "On the circularity of a complex random variable," *IEEE Signal Process. Lett.*, vol. 15, pp. 841–844, 2008. DOI: 10.1109/LSP.2008.2005050 5, 6, 19

[37] B. Picinbono, "On circularity," *IEEE Trans. Signal Process.*, vol. 42, pp. 3473–3482, Dec. 1994. DOI: 10.1109/78.340781 18

[38] B. Picinbono and P. Chevalier, "Widely linear estimation with complex data," *IEEE Trans. Signal Process.*, vol. 43, pp. 2030–2033, Aug. 1995. DOI: 10.1109/78.403373 18, 24, 35, 65

[39] E. Plourde and B. Champagne, "Generalized Bayesian estimators of the spectral amplitude for speech enhancement," *IEEE Signal Process. Lett.*, vol. 16, pp. 485–488, June 2009. DOI: 10.1109/LSP.2009.2018225 2

[40] T. Rohdenburg, V. Hohmann, and B. Kollmeir, "Objective perceptual quality measures for the evaluation of noise reduction schemes," in *Proc. IWAENC*, 2005, pp. 169–172. 87

[41] M. R. Schroeder, U.S. Patent No. 3,180,936, filed Dec. 1, 1960, issued Apr. 27, 1965. 1

[42] M. R. Schroeder, U.S. Patent No. 3,403,224, filed May 28, 1965, issued Sept. 24, 1968. 1

[43] M. Souden, J. Benesty, and S. Affes, "On the global output SNR of the parameterized frequency-domain multichannel noise reduction Wiener filter," *IEEE Signal Process. Lett.*, vol. 17, pp. 425-428, May 2010. DOI: 10.1109/LSP.2010.2042520 48, 60, 73

[44] P. Vary and R. Martin, *Digital Speech Transmission: Enhancement, Coding and Error Concealment*. Chichester, England: John Wiley & Sons Ltd, 2006. 5

[45] R. Vetter, "Single channel speech enhancement using MDL-based subspace approach in Bark domain," in *Proc. IEEE ICASSP*, 2001, pp. 641–644. DOI: 10.1109/ICASSP.2001.940913 32

[46] N. Virag, "Single channel speech enhancement based on masking properties of the human auditory system," *IEEE Trans. Speech, Audio Process.*, vol. 7, pp. 126–137, Mar. 1999. DOI: 10.1109/89.748118 32

[47] M. R. Weiss, E. Aschkenasy, and T. W. Parsons, "Processing speech signals to attenuate interference," in *Proc. IEEE Symposium on Speech Recognition*, Carnegie-Mellon Univ., 1974, pp. 292–295. 1

[48] P. J. Wolfe and S. J. Godsill, "Efficient alternatives to the Ephraim-Malah suppression rule for audio signal enhancement," *EURASIP J. Applied Signal Process., Special Issue: Digital Audio for Multimedia Communications*, pp. 1043–1051, Sept. 2003. DOI: 10.1155/S1110865703304111 2

[49] C. H. You, S. N. Koh, and and S. Rahardja, "β-order MMSE spectral amplitude estimation for speech enhancement," *IEEE Trans. Speech, Audio Process.*, vol. 13, pp. 475–486, July 2005. DOI: 10.1109/TSA.2005.848883 2

[50] ITU-T Rec. P.862, "Perceptual evaluation of speech quality (PESQ), an objective method for end-to-end speech quality assessment of narrowband telephone networks and speech codecs," Feb. 2001. 78

[51] ITU-T Rec. P.862.1, "Mapping function for transforming P.862 raw result scores to MOS-LQO," Nov. 2003. 87

Authors' Biographies

JACOB BENESTY

Jacob Benesty was born in 1963. He received a Master degree in microwaves from Pierre & Marie Curie University, France, in 1987, and a Ph.D. degree in control and signal processing from Orsay University, France, in April 1991. During his Ph.D. (from Nov. 1989 to Apr. 1991), he worked on adaptive filters and fast algorithms at the Centre National d'Etudes des Telecomunications (CNET), Paris, France. From January 1994 to July 1995, he worked at Telecom Paris University on multichannel adaptive filters and acoustic echo cancellation. From October 1995 to May 2003, he was first a Consultant and then a Member of the Technical Staff at Bell Laboratories, Murray Hill, NJ, USA. In May 2003, he joined the University of Quebec, INRS-EMT, in Montreal, Quebec, Canada, as a Professor. His research interests are in signal processing, acoustic signal processing, and multimedia communications. He is the inventor of many important technologies. In particular, he was the lead researcher at Bell Labs who conceived and designed the world-first, real-time, hands-free, full-duplex stereophonic teleconferencing system. Also, he and Tomas Gaensler conceived and designed the world-first, PC-based, multi-party hands-free, full-duplex stereo conferencing system over IP networks. He is the editor of the book series: *Springer Topics in Signal Processing*. He was the co-chair of the 1999 International Workshop on Acoustic Echo and Noise Control and the general co-chair of the 2009 IEEE Workshop on Applications of Signal Processing to Audio and Acoustics. He was a member of the IEEE Signal Processing Society Technical Committee on Audio and Electroacoustics and a member of the editorial board of the EURASIP Journal on Applied Signal Processing. He is the recipient, with Morgan and Sondhi, of the IEEE Signal Processing Society 2001 Best Paper Award. He is the recipient, with Chen, Huang, and Doclo, of the IEEE Signal Processing Society 2008 Best Paper Award. He is also the co-author of a paper for which Y. Huang received the IEEE Signal Processing Society 2002 Young Author Best Paper Award. In 2010, he received the Gheorghe Cartianu Award from the Romanian Academy. He has co-authored and co-edited/co-authored many books in the area of acoustic signal processing. He is also the editor-in-chief of the reference *Springer Handbook of Speech Processing* (Berlin: Springer-Verlag, 2007).

YITENG HUANG

Yiteng Huang received his B.S. degree from the Tsinghua University, Beijing, China, in 1994 and the M.S. and Ph.D. degrees from the Georgia Institute of Technology (Georgia Tech), Atlanta, in 1998 and 2001, respectively, all in electrical and computer engineering. From March 2001 to January 2008, he was a Member of Technical Staff at Bell Laboratories, Murray Hill, NJ. In January 2008, he founded the WeVoice, Inc., in Bridgewater, New Jersey and served as its CTO. His current research interests are in acoustic signal processing, multimedia communications, and wireless sensor networks. Dr. Huang served as an Associate Editor for the EURASIP Journal on Applied Signal Processing from 2004 and 2008 and for the IEEE Signal Processing Letters from 2002 to 2005. He served as a technical Co-Chair of the 2005 Joint Workshop on Hands-Free Speech Communication and Microphone Array and the 2009 IEEE Workshop on Applications of Signal Processing to Audio and Acoustics. He is a coeditor/coauthor of seven books in the area of acoustic signal processing. He received the 2008 Best Paper Award and the 2002 Young Author Best Paper Award from the IEEE Signal Processing Society, the 2000-2001 Outstanding Graduate Teaching Assistant Award from the School Electrical and Computer Engineering, Georgia Tech, the 2000 Outstanding Research Award from the Center of Signal and Image Processing, Georgia Tech, and the 1997-1998 Colonel Oscar P. Cleaver Outstanding Graduate Student Award from the School of Electrical and Computer Engineering, Georgia Tech.

Index